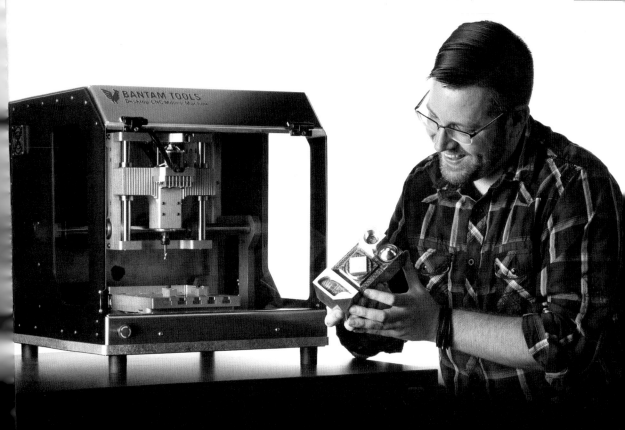

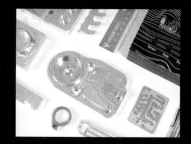

Make: 75

CONTENTS

24

28

52

ON THE COVER:
Nick Seward's one-of-a-kind 3D printer "GUS Simpson."
Photo credit: Jennifer Seward

74

68

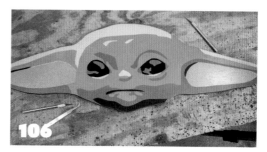
106

STATEMENT OF OWNERSHIP, MANAGEMENT AND CIRCULATION (required by Act of August 12, 1970: Section 3685, Title 39, United States Code). 1. MAKE Magazine 2. (ISSN: 1556-2336) 3. Filing date: 10/1/2020. 4. Issue frequency: Quarterly. 5. Number of issues published annually: 4. 6. The annual subscription price is 34.95. 7. Complete mailing address of known office of publication: Make Community, LLC 150 Todd Road Ste. 200, Santa Rosa, CA 95407. Contact person: Kolin Rankin. Telephone: 305-859-0063 8. Complete mailing address of headquarters or general business office of publisher: Make Community, LLC 150 Todd Road Ste. 200, Santa Rosa, CA 95407. 9. Full names and complete mailing addresses of publisher, editor, and managing editor. Publisher, Todd Sotkiewicz, Make Community, LLC. 150 Todd Road Ste. 200, Editor, Mike Senese, Make Community, LLC, 150 Todd Road Ste. 200, Santa Rosa, CA 95407. Managing Editor, N/A, Make Community, LLC. 150 Todd Road Ste. 200, Santa Rosa, CA 95407. 10. Owner: Make Community, LLC; 150 Todd Road Ste. 200, Santa Rosa, CA 95407. 11. Known bondholders, mortgages, and other security holders owning or holding 1 percent of more of total amount of bonds, mortgages or other securities: None. 12. Tax status: Has Not Changed During Preceding 12 Months. 13. Publisher title: MAKE Magazine. 14. Issue date for circulation data below: Fall 2020. 15. The extent and nature of circulation: A. Total number of copies printed (Net press run). Average number of copies each issue during preceding 12 months:58,620. Actual number of copies of single issue published nearest to filing date: 54,980. B. Paid circulation. 1. Mailed outside-county paid subscriptions. Average number of copies each issue during the preceding 12 months: 43,030. Actual number of copies of single issue published nearest to filing date: 39,838. 2. Mailed in-county paid subscriptions. Average number of copies each issue during the preceding 12 months: 0. Actual number of copies of single issue published nearest to filing date:0. 3. Sales through dealers and carriers, street vendors and counter sales. Average number of copies each issue during the preceding 12 months: 4,450. Actual number of copies of single issue published nearest to filing date: 4,840. 4. Paid distribution through other classes mailed through the USPS. Average number of copies each issue during the preceding 12 months: 0. Actual number of copies of single issue published nearest to filing date: 0. C. Total paid distribution. Average number of copies each issue during preceding 12 months: 47,479. Actual number of copies of single issue published nearest to filing date:44,678. D. Free or nominal rate distribution (by mail and outside mail). 1. Free or nominal Outside-County. Average number of copies each issue during the preceding 12 months:164. Number of copies of single issue published nearest to filing date: 1. 2. Free or nominal rate in-county copies. Average number of copies each issue during the preceding 12 months: 0. Number of copies of single issue published nearest to filing date: 0. 3. Free or nominal rate copies mailed at other Classes through the USPS. Average number of copies each issue during preceding 12 months: 0. Number of copies of single issue published nearest to filing date: 0. 4. Free or nominal rate distribution outside the mail. Average number of copies each issue during preceding 12 months: 839. Number of copies of single issue published nearest to filing date: 888. E. Total free or nominal rate distribution. Average number of copies each issue during preceding 12 months: 1008. Actual number of copies of single issue published nearest to filing date: 889. F. Total free distribution (sum of 15c and 15e). Average number of copies each issue during preceding 12 months: 48,481. Actual number of copies of single issue published nearest to filing date: 45,567. G. Copies not Distributed. Average number of copies each issue during preceding 12 months: 10,139. Actual number of copies of single issue published nearest to filing date: 9,323. H. Total (sum of 15f and 15g). Average number of copies each issue during preceding 12 months: 58,620. Actual number of copies of single issue published nearest to filing: 54,980. I. Percent paid. Average percent of copies paid for the preceding 12 months: 97.93% Actual percent of copies paid for the preceding 12 months: 98.05%. 16. Electronic Copy Circulation: A. Paid Electronic Copies. Average number of copies each issue during preceding 12 months: 3,709. Actual number of copies of single issue published nearest to filing date: 3,384. B. Total Paid Print Copies (Line 15c) + Paid Electronic Copies (Line 16a). Average number of copies each issue during preceding 12 months: 51,189. Actual number of copies of single issue published nearest to filing date: 48,062. C. Total Print Distribution (Line 15f) + Paid Electronic Copies (Line 16a). Average number of copies each issue during preceding 12 months: 52,190. Actual number of copies of single issue published nearest to filing date: 48,951. D. Percent Paid (Both Print & Electronic Copies) (16b divided by 16c x 100). Average number of copies each issue during preceding 12 months: 98.00%. Actual number of copies of single issue published nearest to filing date: 98.18%. I certify that 50% of all distributed copies (electronic and print) are paid above nominal price: Yes. Report circulation on PS Form 3526-X worksheet 17. Publication of statement of ownership will be printed in the Winter 2020 issue of the publication. 18. Signature and title of editor, publisher, business manager, or owner: Todd Sotkiewicz, Business Manager. I certify that all information furnished on this form is true and complete. I understand that anyone who furnishes false or misleading information on this form or who omits material or information requested on the form may be subject to criminal sanction and civil actions.

Adobe Stock-lembergvector, Jennifer Seward, Cody Goddard, Joey Castillo, Tristan Copley Smith, Wesley Treat

Make:

®

PRESIDENT
Dale Dougherty
dale@make.co

VP, PARTNERSHIPS
Todd Sotkiewicz
todd@make.co

EDITORIAL

EXECUTIVE EDITOR
Mike Senese
mike@make.co

SENIOR EDITORS
Keith Hammond
keith@make.co

Caleb Kraft
caleb@make.co

PRODUCTION MANAGER
Craig Couden

CONTRIBUTING EDITOR
William Gurstelle

CONTRIBUTING WRITERS
Heinz Behling, Sarah Boisvert, Gareth Branwyn, Joey Castillo, Daniel Connell, Beck Dalton, Aaron Dietzen, Evan Diewald, Greg Gilman, Agnieszka Golda, Bob Knetzger, Dominik Laa, Tom Lauerman, Jo Law, Helen Leigh, Samer Najia, Niq Oltman, Allen Pan, Nathan Pirhalla, Billie Ruben, Howard Sandroff, Antonio Scala, Wesley Treat, Michael Weinberg

CONTRIBUTING ARTIST
Billie Ruben

MAKE.CO

ENGINEERING MANAGER
Alicia Williams

WEB APPLICATION DEVELOPER
Rio Roth-Barreiro

BOOKS

BOOKS EDITOR
Patrick DiJusto

DESIGN

CREATIVE DIRECTOR
Juliann Brown

GLOBAL MAKER FAIRE

MANAGING DIRECTOR, GLOBAL MAKER FAIRE
Katie D. Kunde

MAKER RELATIONS
Sianna Alcorn

GLOBAL LICENSING
Jennifer Blakeslee

MARKETING

DIRECTOR OF MARKETING
Gillian Mutti

LEARNING LABS

DIRECTOR OF LEARNING
Nancy Otero

OPERATIONS

ADMINISTRATIVE MANAGER
Cathy Shanahan

ACCOUNTING MANAGER
Kelly Marshall

OPERATIONS MANAGER & MAKER SHED
Rob Bullington

PUBLISHED BY

MAKE COMMUNITY, LLC
Dale Dougherty

Comments may be sent to:
editor@makezine.com

Visit us online:
make.co

Follow us:
🐦 @make @makerfaire @makershed
📘 makemagazine
📷 makemagazine
▶ makemagazine
📺 twitch.tv/make
Ⓟ makemagazine

Manage your account online, including change of address:
makezine.com/account
866-289-8847 toll-free in U.S. and Canada
818-487-2037,
5 a.m.–5 p.m., PST
cs@readerservices.makezine.com

Make:
Community

Support for the publication of *Make:* magazine is made possible in part by the members of Make: Community. Join us at make.co.

CONTRIBUTORS

Sarah Boisvert
Santa Fe, New Mexico
(Fabricating the Future of Work)
Host a big hands-on digital badge micro-certification class here in Santa Fe, New Mexico with people collaborating, ideating, and innovating!

Joey Castillo
Brooklyn, New York
(The Open Book and the E-Book FeatherWing)
I would invite a bunch of friends to a giant potluck dinner at my house.

Nancy Otero
Oakland, California
(Transformative Learning)
I'd love to go dancing. I miss taking African dance classes and going to dance salsa.

Issue No. 75, Winter 2020. *Make:* (ISSN 1556-2336) is published quarterly by Make Community, LLC, in the months of February, May, Aug, and Nov. Make Community is located at 150 Todd Road, Suite 200, Santa Rosa, CA 95407. SUBSCRIPTIONS: Send all subscription requests to *Make:*, P.O. Box 17046, North Hollywood, CA 91615-9588 or subscribe online at makezine.com/offer or via phone at (866) 289-8847 (U.S. and Canada); all other countries call (818) 487-2037. Subscriptions are available for $34.99 for 1 year (4 issues) in the United States; in Canada: $43.99 USD; all other countries: $49.99 USD. Periodicals Postage Paid at San Francisco, CA, and at additional mailing offices. POSTMASTER: Send address changes to *Make:*, P.O. Box 17046, North Hollywood, CA 91615-9588. Canada Post Publications Mail Agreement Number 41129568. CANADA POSTMASTER: Send address changes to: Make Community, PO Box 456, Niagara Falls, ON L2E 6V2

PRINTED WITH
SOY INK

Maker at Work

by Dale Dougherty, *Make:* President

Sometimes I have heard said disparagingly: "Makers are just hobbyists" as though what they do is insignificant. But one need only look at efforts of the maker community to produce PPE during Covid-19 to see a rapid response made possible by the combination of skills, mindset and networks of makers.

There are many facets to the term "maker," which I've regarded as its strength. The term "maker' is versatile, just like the people who use it. Making can be play and it can be work; and for the truly lucky, it is both. Also, what you learn to do at work can be applied to your personal life, and vice versa.

Several years ago on a trip to Shanghai, I was introduced to a man who was described to me as an artist, photographer, maker, gallery owner, entrepreneur. "He's a slash man," said Kevin Lau about the man. Kevin has organized Maker Faires in Shenzhen and Xi'an and this man was a friend of his. I had to ask him: "What's a slash man?" Kevin replied that "he describes himself as an artist/photographer/maker/gallery owner/entrepreneur," with a downward slash of the hand emphasizing the transition between each term. Slash man had become so popular a term in China that people were offering workshops on the idea.

While a slashline may be used to indicate all the different facets of a person, it also suggests that the facets are somehow connected, as creativity is a connection between artist and photographer, but also entrepreneur. The slashline describes how we see our value in the world, a short-form resume of who we are and what we do.

In this issue's lead article, "Fabricating the Future of Work" (page 24), Sarah Boisvert writes that makers who are learning digital fabrication are preparing themselves for jobs that will be in demand even more in the future. She says that "employers are re-examining the need for a college degree in many professions." What employers want is people who demonstrate they have the know-how to do the work, and the ability to learn on the job.

At Make: Community, we have a new initiative called Make: Learning Labs, an alternative educational program for older students (17+) who are not enrolled in college or who are unemployed or underemployed, often because of Covid-19. Nancy Otero, who joined us this summer as Director of Learning, writes about the transformative power of making for learners, even those who never heard the term "maker." She has developed a 4–6 month program for those who can learn to develop the skills and mindset of makers by creating and collaborating on projects. Read more about it on page 18.

Gary Bolles, whose father wrote the bestselling book *What Color Is Your Parachute* (the first-of-its-kind book on choosing a career path), told me recently that a young person needs to find the intersection between what they're interested in and what they're good at. Unfortunately, our schools don't do a good job at helping young people find this out.

Introducing young people to the practice of making can help them discover what they like to do and inspire them to see all that they can do. ◑

MADE ON EARTH

Backyard builds from around the globe

Know a project that would be perfect for Made on Earth?
Let us know: *editor@makezine.com*

REMEMBERING RALPH

CONCEPTREALIZATIONS.COM

When Ralph Baer, creator of the earliest home gaming consoles, passed away in 2014, his family raised funds to produce a life-size sculpture commemorating his role as "father of video games" in his hometown of Manchester, NH. 3D artist **Cosmo Wenman** and his company Concept Realizations stepped in to help design and produce a bronze statue more quickly and affordably than traditional routes would offer.

Wenman terms his sculpture workflow "digital waxworks." In it, he designs a piece in software (he uses Blender), then virtually splits the complete model into parts while adding necessary sprues, joints, and flanges for the actual casting and welding. The pieces are then 3D printed on Voxeljet printers using a wax-like material that can be cast with metal.

"It lets the foundry entirely skip the mold making and wax casting processes, which can be extremely time consuming and expensive, particularly on large and complex sculptures," Wenman says. "It's possible to make a life-size bronze from scratch in a matter of weeks." He says it also allows him to design consistent, or even strategic, wall thicknesses, instead of the traditionally uneven hollowness of a sculpture.

The foundries then handle the final 3D prints just like conventional waxes with their normal process. "The actual burnout, casting, and chasing workflow would be largely recognizable to a foundry worker hundreds, or maybe thousand of years ago," he says.

The foundry team installed the sculpture in a local riverside park, renamed in Baer's honor, to fanfare. Wenman, who's been active in 3D printing and scanning since 2009, is meanwhile looking next to the classics.

"I'm using French freedom of information law to compel the Rodin Museum in Paris to make its unpublished 3D scans of Auguste Rodin's sculptures accessible to the public," Wenman says. (All of Rodin's works are in the public domain.) "By this time next year, I hope to be using this process to make high-fidelity, large-scale bronze reproductions of *The Thinker*, *The Age of Bronze*, *The Gates of Hell*." —*Mike Senese*

Jeff Normandin @exitthenorm, Cosmo Wenman

BOOKSHELF WITH A SECRET

YOUTU.BE/N4ZPKUMC29W

Book lovers read to be transported to imaginary new worlds, but these makers built an actual world between their books. A recreation of *Harry Potter*'s Diagon Alley lives on the shelf of Norway-based DIY YouTubers **Nerdforge** (youtube.com/nerdforge). The couple, **Martina** (25) and **Hansi** (27), has been producing DIY content online for four and a half years. Their channel includes large-scale projects such as a camper van conversion series, but they also create art, props, or dioramas inspired by fantasy games and fiction.

For this build, "I discovered the concept of 'book nooks' and I immediately knew I wanted to make something from the *Harry Potter* universe," says Martina. Book nooks are diorama bookshelf inserts and Nerdforge's required forming styrofoam brick walls, laser cut storefronts and glass windows. They also had to solder electronics, program the lights, paint, and glue the finishing touches. Most skills they learned from watching YouTube; however, Hansi, a software engineer, did provide the code for the Arduino that controls the stores' lights. "It might be one of the most insignificant details in the build," commented Martina, "but the top floor [lights] turn on and off at random! It makes the whole piece more alive as if there are actual tiny people living there." These small details are what makes their build so enchanting — flashes and flickers illuminate like magic from Olivander's wand shop, and hand-painted, weathered buildings capture Diagon Alley's atmosphere.

The response has been overwhelmingly positive on the channel with some makers having recreated the build for themselves. The team "are makers at heart" assuring "it will be a long time coming before [they] run out of things to make," and they encourage anyone to DIY their projects. —*Sophie Martinez*

Nerdforge

Congratulations!!

★ To the Movable Makey Contest **WINNERS!** ★

Maker: Shawn Britton
Location: Torrington, CT
Medium: Foam

Maker: Jason Tollefson
Location: Orlando, FL
Medium: 3D printed

Maker: Rhett Pimentel
Location: Powell, WY
Medium: 3D printed

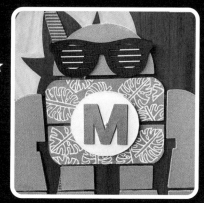

Maker: Lorena Hernandez
Location: Miami, FL
Medium: Paper

Make: Projects

To check out these and other great projects visit makeprojects.com

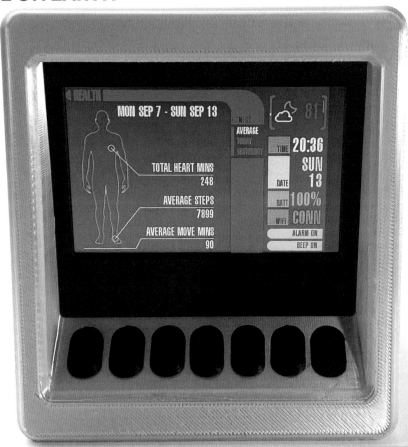

ENTERPRISE DATA TERMINAL

DARIANMAKES.COM/STARTREKDISPLAY

One make leads to the next. **Darian Johnson**'s *Star Trek: The Next Generation*-inspired mini display is the impressive by-product of a shelved attempt to create a smart mirror. "Most of my projects are like most things, you take bits of pieces of earlier projects and build on top of them."

With Covid-19 keeping the Dallas-based cloud technology consultant at home this past June, he turned to Captain Picard for entertainment, but ended up with reinvigorating inspiration to put unfinished business back to work. Johnson's 3D-printed creation not only pays homage to the USS Enterprise's computer interface, it's functional and practical, too. Through APIs and hardware sensors, leveraging the AWS Cloud for data collection, the small countertop terminal provides news, schedule, weather,

indoor temperature, humidity and volatile organic compound strength, and even fitness information. It's the culmination of about six years of challenging himself to make useful hardware outside of the software that keeps him busy at the office.

"I've always been one that learns by building," Johnson says, and he wants to pass on that knowledge. "Almost everything I do, I try to make it as open source as possible." In a perfect world, he imagines building an extensible code base that other science fiction fans can use with the terminal of their choice. But for now, he pressed pause to focus on applying what he learned from this make to yet another: an Afro-futuristic *Black Panther* display, in honor of recently deceased star Chadwick Boseman. —*Greg Gilman*

Darian Johnson

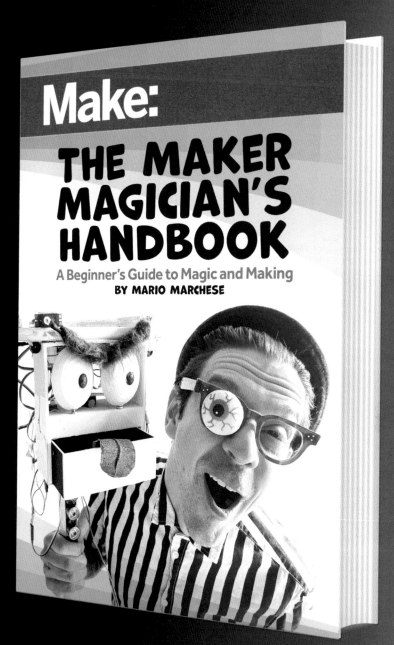

OSHW
Turns 10

Written by
Michael Weinberg

Lessons learned over a decade of **open hardware** endeavors

It is appropriate that, 10 years after the first Open Hardware Summit, open source hardware was a key part of the initial Covid-19 response. Engineers, designers, and medical professionals collaborated from around the world to design and deploy medical equipment to meet the world's unprecedented need.

In many ways, this is exactly what participants had in mind during the first open hardware workshop organized by Ayah Bdeir and held in the Eyebeam art space in October of 2010. They were not the first people to discuss open source hardware — open source activists like Bruce Perens had been advocating for open source hardware since the late 1990s. Nonetheless, that gathering helped lay the groundwork for the modern open source hardware movement.

The idea of open hardware does not exist in a void. It builds on decades of engineering, legal, and cultural work by the open source software community. In fact, most of the structures of the open source hardware community started as structures in the open source software community. While many of those central tenants remain the same, a decade of applying software's ideas of openness to hardware has created a culture all its own.

By 2020, the Open Hardware Summit (virtual this time thanks to Covid) had grown into an international event, bridging together a community spread around the world.

Why 2010?

By 2010, two related trends began to converge. The first was the arrival of "good enough" hardware. Although things like processing power continue to increase rapidly, by 2010 hardware components did not need to be on the absolute cutting edge in order to do genuinely interesting and useful things. As articulated by Bunnie Huang at the 2011 Open Hardware Summit, this dynamic made it relatively easy for small businesses and groups of people to create compelling hardware without having access to multi-million dollar research pipelines.

This relative ease of creation helped spur the second trend: the emergence of a critical mass of companies and communities creating accessible, open hardware. Adafruit, Arduino,

MICHAEL WEINBERG is the president of the board of the Open Source Hardware Association (OSHWA) and the Executive Director of the Engelberg Center on Innovation Law & Policy at NYU Law. Find him online @MWeinberg2D and michaelweinberg.org.

Attendees pose outside the 2012 OSHW Summit.

OpenBCI demos their headset at the 2018 Summit.

Public domain, Jacobgibb.com, Open Hardware Summit

Evil Mad Scientist Laboratories, Makerbot, RepRap, SparkFun — by 2010 these efforts were not isolated incidents. They were a budding community that validated each other.

That community quickly began to formalize itself. That initial workshop was quickly followed by a number of important milestones, including kicking off an annual Open Hardware Summit, creating an open hardware definition, agreeing on a logo, and, led by Alicia Gibb, establishing the Open Source Hardware Association (OSHWA) to house it all. A few years later, the Gathering for Open Science Hardware (GOSH) created a manifesto specifically for bringing open source hardware to the scientific community. All of this happened in collaboration and dialogue with the larger maker movement, which was also growing.

Growth and Challenges

The needs of the open hardware community increased as more people joined. Once the community grew beyond a relatively small group of people with in-person connections, Phil Torrone realized that writing down the unspoken

Participants in the first open hardware workshop, 2010.

rules of open source hardware would make it easier for new people to join the community. Documenting the rules acted as an invitation to new community members, giving them confidence to navigate the collective expectations of open source hardware.

This period also helped to show that open source hardware theories also worked in practice. In a prelude to today's Covid responses, the Safecast radiation sensor project organized radiation level tracking in response to the Fukushima Daiichi Nuclear Power Plant disaster. Open source hardware companies multiplied across a wide range of industries. While there were high profile stumbles — such as the flagship open source hardware company Makerbot going closed — the trend in open source hardware was towards growth and new applications.

That growth brought additional challenges. Although OSHWA maintained the community-created definition of open source hardware, no one owned the term "open source hardware." The celebrated "open gear" open source hardware logo was similarly free from any one individual or organization's control. While this openness brought a number of benefits, it also meant that nothing prevented decidedly not-open hardware from advertising itself as if it was open. This behavior — sometimes described as "open-washing" — threatened to undermine the term open source hardware and render it meaningless.

In response, OSHWA decided to create a new open source hardware certification program and certification logo. The free program gave open source hardware creators and users an easy way to identify open source hardware that met the requirements of the open source hardware definition. Regardless of how a piece of hardware was advertised, a certification logo meant that it complied with the community definition of open

source hardware.

The certification program also gave OSHWA an opportunity to consolidate information about one of the other perpetual open source hardware challenges — licensing.

Licensing is one of the biggest differences between open source software and open source hardware. Software is "born closed" — automatically protected by copyright from the moment it is written. A piece of software is fully protected by copyright, meaning that anyone who wants to use it needs permission from the creator — a license. Over the decades, the open source software movement has capitalized on the born closed nature of software, using licenses to spread the requirements of openness beyond people with an inherent interest in openness.

In contrast, major parts of hardware are "born open" — not automatically protected by copyright or any other kind of right. While some *parts* of hardware may be protected by copyright, other parts may be free by default. This creates a much more complicated rights situation, making it much harder to understand when a license is necessary — and when a license can require other users to be open.

Although existing open source software licenses can be used to license portions of open source hardware, the community also created licenses drafted with the specifics of hardware in mind. Various licenses, such as the TAPR Open Hardware License, the Solderpad Open Hardware License, and the CERN Open Hardware Licenses emerged as options for the community. While these licenses do not necessarily clarify when a piece of hardware requires a license in the first place, they can help give the community confidence that — to the extent that they are necessary — the licenses will perform as expected. CERN's recently released second generation licenses use "flavor of openness" designations to help make navigation even easier.

Open Source Hardware in 2020

Ten years in, the open source hardware community continues to grow. OSHWA's certification program includes hardware from over 40 countries on five continents. Open Hardware Month activities in October include a similarly international set of events. GOSH continues to help spread open source hardware in the international science community.

The global response to Covid vividly illustrates the importance of open source hardware approaches. Teams from around the world came together to rapidly create, innovate, and distribute a broad range of medical supplies to communities that needed them most. Their open approach allowed improvements and best practices to propagate quickly, and for communities to easily modify equipment as needed.

If 2010's original open source hardware workshop was about exploring a theory of open source hardware, 2020's open source hardware community proves that theory out every day.

The Next 10 Years

Open source hardware is all about collaborative innovation, so the next 10 years will look very different from the first 10. While we cannot anticipate all of the challenges, some opportunities are clear:

Marking the path for open source hardware success — There are scores of examples of successful open source hardware companies. While they are beginning to highlight common factors for success, we are far from a playbook (or playbooks) for successfully creating open source hardware. Further distilling the lessons for open source hardware success will make it even easier for a broader open source hardware community to succeed.

Diversify open source hardware — Although the open source hardware community is already an international one, it will continue to work to be a community that welcomes and celebrates members from a broad range of backgrounds and experiences. In addition to individual diversity, the open source hardware community will also

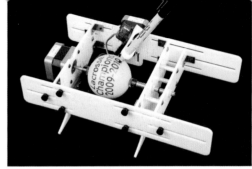

Evil Mad Scientists' open-source EggBot prototype.

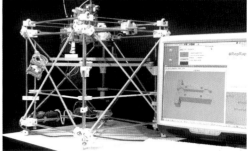

The RepRap movement established an open source trend with the 3D printing community.

Ted Ullrich, Wildell Oskay, Adrian Bowyer

work to incorporate more types of hardware and hardware applications.

Easier academic paths — Some of open source hardware's strongest advocates are in academia. Unfortunately, it can be hard for traditional academic structures to recognize contributions to open source hardware. The academic portions of the open source hardware community continue to work to make sure that contributions to open source hardware are valued equally with contributions to less open projects.

More open components — One of open source software's great strengths is that any given piece of open source software is built upon a number of open source libraries and other building blocks. The open source hardware community will work to build more open components, allowing open source hardware practices to extend deeper into the hardware world.

Keep growing the community — The open source hardware community has grown in the last 10 years, but there is plenty of room to keep going. As open source hardware becomes more common and accessible, the community will continue to expand, finding (and building) new ways to use open source hardware. ✪

Transformative
Written by Nancy Otero
Learning

We're launching
Make: Learning Labs

NANCY OTERO is the Director of Learning at Make:, and is a fan of laser cutters and AI. She has spent the last 10 years working with educational organizations designing environments that mix technology with project-based learning.

Project from FAB! — a lamp to help small kids fall asleep.

For humans, making is as natural as breathing. There is a history of 3.3 million years of making, told by the oldest hominid-made artifact we know. For my father, making is his way to relate to the world. So fixing, improving, creating, and exploring was how he would spend his time at home during my childhood. Even though I wasn't invited to as many projects as my brother was, I participated in enough of them to see the world as something that could be made. I could imagine cutting holes or adding electronics to the stuff around me as a default nature. I didn't have to choose making; it was mandatory in my family.

Students at Portfolio School, NYC

Nancy Otero, Doug Schachtel

Making showcases not just what you know but what can you do with what you know.

Knowing the value of making, I was very excited to join *Make:* this year. For several months I have been working on developing **Make: Learning Labs**, a modular program designed for small groups of young people 16 and older. This is a program model that we will share and hope to have it replicated by partners. The program provides an alternative to traditional education programs that helps young people discover what they are interested in and what they can do. It will encourage students to experiment with new ideas and new skills. The program can help them become good learners who can work in a creative and collaborative environment.

The curriculum has an online component that can be done at any time that is convenient for the participant, and an in-person lab component where participants work together and individually on different platforms to create projects (though these components can run online during Covid).

Here's why this program is so important now:

- According to a recent survey done by ECMC of high-schoolers, 74% of the Z generation believe a skills-based education (e.g., trade skills, nursing, STEM, etc.) makes sense in today's world, and 63% said that the top place to learn is a hands-on lab.
- As an example, NASA's recruitment creativity test was used in a longitudinal study. 98% of children 4–5 years old score at the highest of the test, something called creative genius. By the time children were 10 years old, 30% scored those points; by 15 years old, only 12%. Just 2% of adults are creative geniuses. The more schooling, the less creative.
- Today 41% of teens haven't attended an online or virtual class since in-person school was canceled, and the International Labour Organization projects 273 million young people in NEET (Not in Employment, Education, or Training) in 2021.

Without a viable alternative, many young people are at risk of dropping out, taking a path where the opportunities for learning are fewer, and limiting their career in the long term. Make: Learning Labs is a program that can engage young people and develop agency around the world.

How Make: Learning Labs Works

This new initiative provides a learning trajectory to help participants gain confidence in their own ability to think and create. The program has

Participants of the Makers in Residence program in Mexico.

three phases: 1) immersion in making; 2) a deep dive to develop skills in software, electronics, and digital fabrication; and 3) independent projects that applies the skills and knowledge to culturally relevant real-world problems. Participants will learn to use a problem-solving framework, such as design thinking, and project management tools for agile methodologies such as Slack, Github, and Asana. Participants will meet with mentors to talk about their interests, and develop a custom learning plan for the next module.

Throughout the program, students will be expected to develop projects, starting small and increasing in complexity, both on their own and in collaboration with peers. The program will culminate in a capstone project that introduces the process of innovation. It follows a process that involves need-finding, contextualization, prototyping of several solutions, and iterative testing. This phase culminates with the students sharing a presentation and demonstration of the innovative solution. Students will develop and showcase projects on makeprojects.com. Make: Learning Labs participants will have at least five projects for their portfolio that can be used, if desired, to apply to college or a new job.

We know about the many obstacles that schools and teachers face implementing making. That's why we are developing Make: Learning Labs with five main characteristics:

1. Creative freedom for anyone to design their

FAB! Project: Shocking-alarm clock.
Think twice before you hit snooze!

Nancy Otero, Gabriela Calderon

Learn how to make projects and you'll see implementable solutions everywhere

own activities, and get inspired by, use, or tinker with already posted making activities

2. A research-and-practice-based framework and developmental trajectory for problem-solving and project-management that binds together the making activities

3. A culturally relevant compass on the activities and pedagogy that guides the creation and implementation of making activities for inclusion

4. Professional development for facilitators and coordinators to implement the program

5. A platform to post projects, collaborate, and create a learning community

When we make, our flow of thought answers to the changes and feedback of the materials, tools and projects which we work with. These tools and materials think in us, as we think through them. Learning how to use scissors makes the world cuttable; learning to code makes the world computable. Learn how to make projects and you'll see implementable solutions everywhere. We want younger people to think about the future and be able to make those thoughts a reality. Anyone can become a maker and any maker an innovator. The Maker Community rolled their sleeves and helped make supplies for hospital and essential works. This is a call for the Maker Community to help again, to support our young people to find their purpose and path to the future.

You can access more information about the Make: Learning Labs at learn.make.co. We are

Teachers at Beam Center learning robotics

looking for partners that want to implement the program in their school, makerspaces, community centers, etc. Those interested can email me at learninglabs@make.co.

Launching the Labs

I've been combining making and learning for years, in both academic and professional settings, and I bring this wealth of experience and expertise to the Make: Learning Labs.

In 2012 as a grad student at Stanford University, through the magical Transformative Learning Technology Lab, I discovered educators that link making and learning, I was drawn to it immediately. I brought those lessons with me to Mexico, where, from 2013–2018, I co-founded and led FAB!, a non-profit that worked with high-school students from underserved communities. The goal was to give that group a useful activity to engage with. It ran as an after-school program, a custom program modeled after the Makers in Residence designed by Prof. Paulo Blikstein, who had brought the digital fabrication lab concept to Stanford (where I worked with him) and was the first to adapt it for secondary school instruction in the STEM fields. During the program, participants learned making skills, used design thinking to find a problem they care about, prototyped a solution using their new skills, and shared their project with the community.

After running this project for three years, I learned that these teenagers liked making and they can do incredible projects even without any previous experience or technical background. I witnessed students develop a device that could be installed in cars to measure a driver's cognitive abilities before being permitted to drive. The goal of this team was to avoid car accidents. This group of teens (some of whom only having just accessed a computer in their school lab) interviewed people, programmed an app, and connected the output of the app with

Making gives tangible form to a legacy of human ideas while adding personal meaning

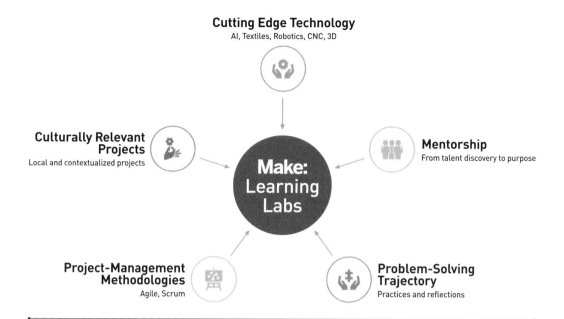

Cutting Edge Technology
AI, Textiles, Robotics, CNC, 3D

Culturally Relevant Projects
Local and contextualized projects

Mentorship
From talent discovery to purpose

Make: Learning Labs

Project-Management Methodologies
Agile, Scrum

Problem-Solving Trajectory
Practices and reflections

a microcontroller to block a car key. I was sold. My next question: How could one amplify the experience with academic learning?

I went on to join the Beam Center in NYC in 2014 with the goal of bringing making to teachers. I created a program for public high-schools' teachers that would help bridge making and academic learning through project-based learning. Teachers partnered with artists and engineers to design big projects where students could apply the curriculum. The program grew and has reached thousands of students and more than 100 teachers. But those activities occurred a few times per year, even when the improvements in students' learning and engagement were visible. I started wondering if making was a complementary activity or its own way of learning. How would it look like for a school to turn all its curriculum into projects?

I had the opportunity to find out when I became the Founding Director of Learning Design and Research at Portfolio School in 2016, a project-based learning independent micro-school in NYC. This was a dream come true. With the mission of redesigning schooling given everything we know today about learning, we eliminated tests, grades, 45 minute schedules, and created a learning framework where making became a tool to apply and internalize powerful human ideas. Through four years of this experience, I can say with certainty that making is a fundamental piece of learning, a piece that gives tangible form to a legacy of human ideas while adding personal meaning. For everyone, since kindergarten.

And now I'm so happy and excited to join Make: as Director or Learning, where I have applied all the lessons of this journey to launch Make: Learning Labs. We've spent months setting it up and look forward to connecting with you about it further. See you there. ◐

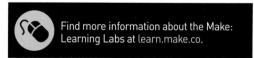

Find more information about the Make: Learning Labs at learn.make.co.

Pilar Perez, Doug Schachtel

FABRICATING THE FUTURE OF WORK

Written by Sarah Boisvert

DIGITAL FABRICATION IS PREPARING MAKERS FOR ENGAGING, WELL-PAYING CAREERS TODAY AND INTO TOMORROW

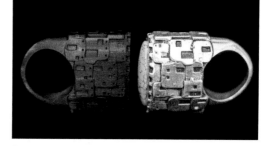

In the early years of the maker movement, who would have predicted that digital fabrication tools could become as affordable and user-friendly as they are today? No longer needing specialists to keep them running, fabrication machines are accepted by mainstream users as tools in their daily lives. CNC routers, laser cutters, digital sewing machines, vinyl cutters, and 3D printers are now found beyond makerspaces and garages, in classrooms, libraries, museums, and startup accelerators.

The resulting democratization of digital fabrication has had an especially powerful impact on entrepreneurs. Suddenly barriers to market entry were drastically lowered and new companies sprang up, offering everything from 3D-printed jewelry to laser-engraved advertising specialties, and drones made from laser-cut parts to CNC-machined home furnishings. Coronavirus accelerated the number of makers who saw opportunity in a local supply chain and there are many stories of high-volume PPE production led by maker-entrepreneurs.

But being a successful entrepreneur is not an easy path and is certainly not the career of choice for everyone. After all, an average of only 20% of U.S. new businesses will still be around beyond the 5-year mark. So, what does digital fabrication hold in store for makers who want an engaging, well-paying career path but don't want to run their own business?

Digital Transformation Across Industries

Just as digital tools have become mainstream in fields such as education, these new technologies are disrupting a range of industries. For the last two years, I have been researching technology's impact on jobs for the Future Workforce Now project led by the National Governors Association, The Fab Foundation, and the international human-development non-profit FHI360. We found that in a short period of time, industries from agriculture to warehousing have undergone digital changes that offer big opportunities for anyone with digital fabrication skills.

Making is no longer the purview of manufacturing alone. One of the most eye-opening comments during the Future Workforce Now project's roundtables came from Walmart:

Ring designed by Mayte Cardenas in CAD, fabricated on a FormLabs SLA 3D printer by Fab Lab Hub, then cast in sterling by Superior Casting.

"Now that we have janitorial robots, we have two concerns. One is what do we do with all of our frontline workers? The other is who is going to program, monitor, and repair the robots?"

The answer lies within the maker movement. Young digital natives are already using the tools that build and repair robots, and have hands-on experience gained in First Robotics competitions and the like. Walmart confirms that automation is happening all around us and the shift from "blue-collar" to what IBM's Ginny Rometty calls "new-collar" jobs is urgent. While we are waiting for the kids to grow up, employers and governments must invest in upskilling today's workers and welcoming adults with non-traditional digital fabrication training.

Bio-printing is often in healthcare news. But other 21st-century technologies are routinely being used in diagnostic labs and operating rooms. Quick, inexpensive tests using "Lab

Walmart's robot janitors will need human builders and technicians with highly technical fabrication skills.

SARAH BOISVERT is the founder of Fab Lab Hub and a co-founder of the New Collar Network, a non-profit issuing digital badges for "New Collar Job" skills. She is the author of *The New Collar Workforce* and *People of the New Collar Workforce*.

"TODAY'S LASER, CNC, AND 3D PRINTING TOOLS PROVIDE CAREER PATHWAYS FOR ANYONE IN OUR COMMUNITY WITH CURIOSITY, DEDICATION, AND A WILLINGNESS TO LEARN." —Potomac Photonics president and CEO Mike Adelstein

A complex microfluidic device manufactured by Potomac Photonics using laser micro-machining.

John Ford, an engineer at Potomac Photonics, a micro-fabrication contract manufacturer. *

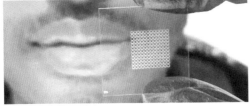

Micro holes machined in glass by Potomac Photonics team.

on a Chip" microfluidic devices are possible because companies like Potomac Photonics, which I co-founded with Dr. Paul Christensen, can laser micro-machine 10- to 30-micron-wide channels in which tiny liquid samples flow. Even traditionally conservative medical device companies like Medtronic have developed robotic surgical assistants in order to increase the accuracy and repeatability of human doctors.

These medical applications have created many types of new-collar jobs that utilize digital fabrication skills, often without a college degree. Explains Potomac Photonics president and CEO Mike Adelstein, "Today's laser, CNC, and 3D printing tools provide career pathways for anyone in our community with curiosity, dedication, and a willingness to learn." A case in point is John Ford, who joined Potomac as a laser operator in 2007 right out of high school. He is now an engineer, and extolls how engaging his job is: "Even on a bad day, how great is it to have a job where you get to play with lasers all day?"

The centuries-old traditions of fashion are also undergoing a digital awakening. In his new venture since departing 3D Systems as CEO, XponentialWorks founder and CEO Avi Reichental and his partner Jenna Jobst created the JenaviGuard, a 3D-knitted face mask for coronavirus protection. Shoe manufacturer New Balance is adding 3D-printed components to shoes now that production equipment can meet volume and material requirements.

Avi points out that the advancement in technology allows for a new human-machine collaboration that opens mankind's potential for creativity. Digital fabrication when incorporated with artificial intelligence and generative design allows humans to do what we do best — innovate!

A Revolution in Hiring Practices

Along with all this new technology comes a revolution that is simultaneously happening in hiring practices. Driven by the skills gap and the urgent need for digital workers, employers are re-examining the need for a college degree in many professions. We've all seen the self-taught programmers who get offered six-figure salaries right out of high school, and that is now a possibility for digital fabrication jobs.

Potomac Photonics, Inc., Formlabs, Jenna Jobst, TiraHowardPhotography.com

Formlabs has teamed up with companies like New Balance to make custom shoe components.

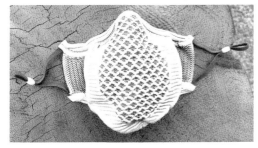

JenaviGuard facemasks, printed at large scale with converted Stoll flat-knitting machines.

Jewelry designer Mayte Cardenas discusses 3D designs with Sarah Boisvert. *

Sarah Boisvert's book *People of the New Collar Workforce* includes augmented-reality videos and interactions. Use the RealityX2 app to scan this cover image and the photos in this article marked with a * for AR previews of its extras.

IBM was one of the earliest companies to hire based on skills, led by the strong beliefs of Ginny Rometty, former CEO and now executive chairman of the digital powerhouse. She has said that as many as 15% of manufacturing jobs at IBM require no college degree and she has been a leader in driving the White House Workforce Task Force. Along with Apple CEO Tim Cook and other industry leaders, the Task Force helped create an executive directive for all federal job hiring practices to be revised in order to focus on skills over degrees. This is impressive since the U.S. Government is the nation's largest employer.

Opportunities for Makers

The World Economic Forum in their 2018 *The Future of Jobs Report* predicts that 75 million jobs will be lost to automation by 2022. On the flip side, they see almost double that number of people, 133 million, will be needed to fill the new kinds of 21st-century jobs that are already emerging. Makers have just the types of skills — digital design, robotics, 3D printing, laser and CNC machining, AI, predictive analytics, and more — that these jobs demand. Further, my research with 200 employers in 2017 showed that employers are hungry for workers with problem-solving experience, something learned naturally in makerspaces, fab labs, and schools teaching with project-based learning methodologies.

The Future Workforce Now toolkit that was developed for states to implement new policy strategies emphasizes the need to foster new Lifelong Learning models. Although 2- or 4-year degrees may not be a digital job requirement, specialized skills are definitely needed. These can be acquired via skill-specific training that is short in duration and affordable, such as the digital badges from the New Collar Network program (newcollarnetwork.com). Even for makers with experience, a micro-credential can add depth to a resume.

It is clear that powerful trends in technology, industry, hiring practices, and education are converging to create a new workplace. With makers' mastery and advancement of digital fabrication tools, we are creating an opening to fabricate this very Future of Work for ourselves. ●

OUTSIDE THE BOX

NICK SEWARD'S EXPERIMENTAL 3D PRINTERS WORK UNLIKE ANYTHING ELSE

Written by Caleb Kraft

A

B

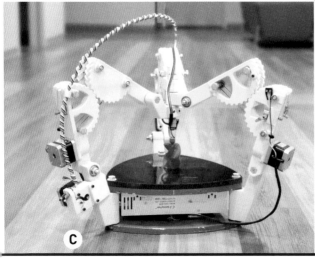

C

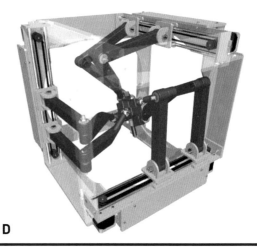

D

At this point, we all have a mental image of a 3D printer: a basic box with a bed and extruder. Even with delta-style machines being somewhat common, the rectangular box concept dominates 3D printing. However, if you've been involved in the RepRap community or been to some of the bigger Maker Faires in the United States, you may have come across a display of interesting printer designs that, in comparison to the squarish Cartesian style, seem bizarre and unique. These are the works of Nick Seward.

Nick Seward, left, with 3D printing pioneers Josef Prusa, center, and Brooke Drumm, right.

Take the Wally (Figure **A**). This printer looks a lot like a cute robot printing onto its own belly. This is largely due to the SCARA (Selective Compliance Assembly Robot Arm) configuration, in which its arms move in an almost humanoid manner in the X- and Y-axes.

Then there's the Wheelios (Figure **B**), which also uses a SCARA arm, but this time mounted on a wheeled, rolling frame. It can, in theory, print indefinitely in one direction.

GUS Simpson is a spider-esque printer with three articulating "legs" that form the motion and the frame (Figure **C**). It's a delta-type machine that undulates and flows in its entirety. The full effect is mesmerizing to watch: you can see it at youtube.com/watch?v=af7XkgxzmXE.

Seward has physically built at least 13 unique designs and variations, and innumerable builds of individual machines. He often shares the full designs to the RepRap community.

Seward's introduction to 3D printing came from exposure to a CNC mill at his first teaching job. He was instantly enamored, and endeavored to create his own. Seeing the fun he was having with this area of exploration, his wife purchased him a gift, one of the original wooden Printrbot 3D printers. A short period of printing the typical tchotchkes left Seward feeling a little empty, and he wanted to give back to the RepRap community.

MORE PECULIAR PRINTERS

Here are a couple innovative machines from other makers that you can build too.

HANGPRINTER

Need to print huge? Forget building an awkward frame, simply run guide strings to the corners of any room, and hang the extruder from the ceiling! The Hangprinter conforms to any room it is in, and can print as big as you'd want to go. Find the build files at hangprinter.org.

WHITE KNIGHT BELT PRINTER

The White Knight's bed is a rotating belt, and the head is tilted, which allows you to print items that extend indefinitely, or at least as far as you provide support. The belt-fed bed also works for continuous printing of multiple items. Build files at github.com/NAK3DDesigns/White-Knight.

Seward's ensuing contributions have been fascinating and numerous. They also tend to have a theme of not being just boring iterations on the common box. Many, such as the Sextuperon (Figure **D** on page 28), require complicated software and electronics that are still being ironed out to this day. So why does he do it?

"At first, I told myself it was to save money and ultimately save others money. I think that was a lie," he says. "I think I just enjoy designing using first principles and breaking conventional wisdom with engineering and science. The 3D printing maker movement is full of garage heroes. A side effect of having a large portion of the community without a formal education is a lot of common-sense knowledge that might not actually be true. I remember getting scoffed at for even proposing the idea of a printer that wouldn't require rails."

3D printers are still just a hobby for Seward. He is currently an instructor at the ASMSA school for the gifted in Arkansas, teaching computer science. But 3D printing finds its way into his classroom, and the results have been significant.

"At ASMSA, I love to get students involved in experimental projects that are at least 3D printing adjacent. The best infill pattern, cubic subdivision, was the brainchild of one of my students," Seward says proudly. "Other software projects of note include a nonplanar slicer (rough prototype at this point), using Steiner trees to reduce support material, and compound bridging."

What does Seward dream about? What big thing would he like to see come to fruition? "Long-term, I want to print a house," he says. "There is no reason a few semi trucks with massive Simpson arms can't roll up to a job site, connect, and print a house. It isn't even a question of economics. Of course it will be cheaper to print a house. The two holdups are regulations and R&D costs. It is just a matter of time before the labor is expensive enough or the R&D cost is low enough to make this happen. Don't get me wrong, there are people printing houses right now. I just want to see it go mainstream."

After watching GUS Simpson in action, we want to see this too. ⊘

Find Nick Seward's machines, design files, and more at nickseward.net.

LEAFHOPPER LEGS

One of the most magical moments of open source collaboration led to the gear arms on GUS. "I had met [RepRap user] Guizmo in person in Texas," says Seward. "He came up from Mexico and I made a trip to meet him and deliver beta parts for GUS Simpson. In the thread at reprap.org/forum/read.php?178,233674, Guizmo put forward what I thought was a horrible idea of Simpson arms. However, I used the 'yes and' approach. After half a dozen back and forths, he posted this (Figure **E**). It gave

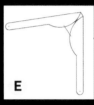

me the 'aha' I needed to get to the gear arms that GUS Simpson uses. This (Figure **F**) is what I had three days later:

E

F

It wasn't that long after that *Science* came out about the leafhopper [insect] having gears like this too: cam.ac.uk/research/news/functioning-mechanical-gears-seen-in-nature-for-the-first-time. An amazing coincidence!"

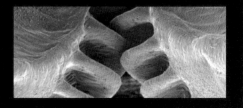

Nick Seward, Malcolm Burrows and Gregory Sutton

CALEB KRAFT is senior editor for *Make:* magazine and has been 3D printing tchotchkes and useful parts since 2013. He often laments the lack of interesting design in mainstream 3D printers.

CNC EVERYTHING!

Talk of *digital fabrication* today is largely focused on 3D printers, laser cutters, and CNC routers and mills; before their rise to prominence, however, some industries relied on a range of other computerized tools.

Those tools have now become available for home workshops, typically quite affordably. They offer a wealth of creative opportunities for makers of all types — perhaps your next project would benefit too. —*Caleb Kraft*

A

A EMBROIDERY MACHINES

Brother and Baby Lock make a series of CNC embroidery machines that are cheaper than $300. You feed a design to the machine and a well-made patch or embroidered design comes out. The software is still daunting, though.

B PEN PLOTTERS

Just like a vinyl cutter but with a pen instead of a blade. Before the inkjet came along and took over, these were everywhere that needed to create blueprints and large-format documents. Now, they fill a niche of computer art that is growing. Find them on Twitter by searching #plottertwitter.

C HOT WIRE FOAM CUTTERS

These systems use a hot wire to slice through blocks of foam. In industrial settings these are invaluable for prototyping and packaging. There are home versions you can find for pretty cheap as well.

D VINYL CUTTERS

A swiveling blade is pulled along material, making precise and accurate cuts. Typically used for sign making and outdoor stickers, but other applications are numerous. Companies like Cricut, Brother, and USCutter continue to release updated versions for crafters.

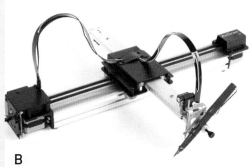

B

C

D

Caleb Kraft, Evil Mad Scientist, Polyshaper, USCutter

THE NEWNESS

DIGITAL FABRICATION MACHINES JUST KEEP GETTING BETTER. HERE ARE SOME OF THE LATEST WE'RE EXCITED FOR

Written by Caleb Kraft

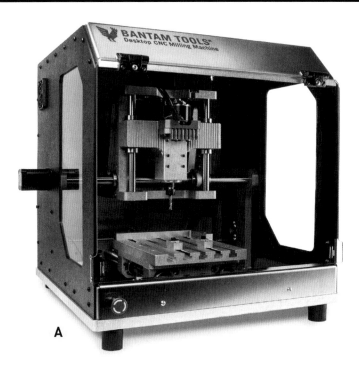

A

Technology is always advancing, and digital fabrication machines are no exception to that march. In recent years we've seen all types of digital build tools improve by leaps and bounds, from mills to routers to 3D printers. With more reliable hardware designs and more intuitive software interfaces, these machines are now easier for anyone to use. We keep our eyes on all the new releases; here are a few that stood out this year in our reviews. See them in action at makezine.com/go/make-workshop.

CNC MILL

A Bantam Desktop CNC Milling Machine
$3,600 bantamtools.com

Bantam's original machine, the Othermill, was primarily designed to make PCBs. Their new CNC mill, however, steps up to get into a bit tougher work. Of course it can still do the soft stuff like plastics and wax, but it's also capable of chewing through aluminum and brass. Sized to fit on a bench, the fully enclosed system would make sense in a small prototype shop or makerspace. Of all the mills in this size and price range, this one is currently the beefiest, with thick guide rails and leadscrews providing great stability.

The software interface is one area where Bantam is attempting to stand out. Instead of using a typical CNC controller package (such as Mach, PathPilot, or LinuxCNC), they've developed their own, aimed at making it easy for beginners to get going. This includes a modern, clean visual style and even some basic automatic path creation based on vector files. Additional features, like advanced edge finding, are available through a paid subscription service (similar to paid add-ons in Fusion 360).

The built-in tool library includes pre-programmed feeds and speeds, which should really help beginners get through the initial learning curve. But, although companies like Bantam are helping advance the ease of use of these tools, you'll find that CNC milling is still a much more complex process than 3D printing or laser cutting.

Our test machine was solid, fully capable of pulling off our projects. I appreciated the enclosure and even the visual design. You could probably build yourself a cheaper mill this size from scratch, but you'd end up investing a lot of time and money to get the fit and finish, as well as features, that the Bantam offers. You'd also be missing out on support.

CNC MILL

B Tormach XS Tech
$3,500 tormach.com

Tormach has established its brand as a solid choice for beginners and small- to medium-sized professional machine shops due to their relatively compact and reasonably priced mills. However, their machines, while much smaller than full production VMCs, are typically still sizable enough that you need to do some room planning and budgeting. Their XS Tech mill breaks that concept by being small enough to fit on a desktop.

The XS Tech is fully enclosed and capable of milling materials as hard as aluminum or brass. It ships with a 10-inch touchscreen and full controller running Tormach's PathPilot software. This is where the XS Tech mill really shines. The learning and transition experience going from this desktop device to Tormach's biggest machine is seamless. They use the same interface (PathPilot), and Tormach even offers a simulation environment that's good for classroom use. A room full of students can run simulations while one project is being milled.

At the price offered, not only would the XS Tech make sense in schools and makerspaces, it makes perfect sense for someone hoping to start a business that needs milling, as your upgrade path will be very nearly seamless as you grow.

The X and Y axes are belt-driven, and the spindle is rated a bit smaller than some others in a similar price range and size, but in our testing it did chomp through aluminum as long as I kept the settings a bit conservative.

B

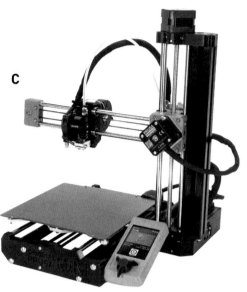

3D PRINTER

c **Prusa Mini 3D Printer**

$350 prusa3d.com

For the last couple years we've seen the price of 3D printers dropping down to under $500 for something that gives you pretty decent results out of the box. Usually there's a caveat that comes with these: "Buy one, and you'll have to tinker a bit, but you can make it print really well." Josef Prusa and his team of engineers must have gotten sick of hearing that because they've released a $350 printer that prints awesome out of the box without any tweaks — the Prusa Mini.

As the name suggests, the Mini is smaller than their flagship i3 MK3S, and so is the price tag. Don't think that they've cut corners in quality or user experience though — the Mini actually comes with some new features that we're eager to see trickle into Prusa's other products.

The color screen interface with full-model previews is really nice, and I'll never stop extolling the benefits of removable flexible print beds. The optional filament sensor ($20) also adds a little peace of mind.

The Prusa Mini has become my first recommendation when I hear people say they're thinking about trying 3D printing. The price is highly competitive and the features don't skimp.

CNC ROUTER

d **Sienci LongMill**

$1,400 as reviewed (largest size with all upgrades) sienci.com

For under $2,000 there are several options for CNC routers available, but not many offer the features of the LongMill by Sienci Labs. This kit has an interesting construction consisting of aluminum brackets, leadscrews, rolling wheels, and 3D-printed connectors. You supply your own wasteboard, which also serves as the base of the whole unit.

Assembly was very easy, resulting in a CNC router that I could (awkwardly) lift and carry around by myself, and was capable of cutting and carving 30"×30", for under $1,500. I really appreciated the leadscrew-driven axes as opposed to the typical, cheaper belt-driven systems. The machine performed quite capably doing basic cuts and even some advanced 3D carving.

Bonus: It may be a niche case, but the fact that you just screw this onto a sheet good means you can also cut *through* the sheet and move the whole router around to do CNC work on items that are just too big to move, like carving sculptures in logs — neat! ●

ONES TO WATCH

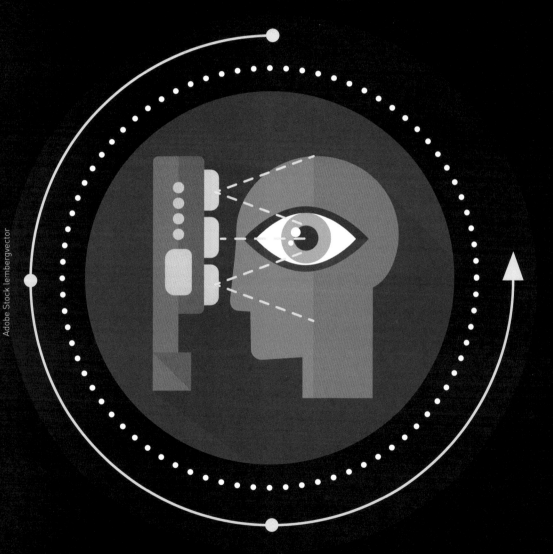

We've heard rumors, seen some demo videos, and even played with some pre-release models. Here are a few upcoming machines that should be launching imminently, possibly before this issue reaches your mailbox.

Peopoly Phenom XXL

Resin printing has seen a few announcements for much larger printers this year, mainly thanks to the cheap LCDs that allow for MSLA (masked SLA) to work. The Peopoly Phenom XXL was just announced and looks to be bigger than many FDM printers, even ones that are considered large. With a print area of 527mm×296mm× 550mm, SLA isn't just for small parts anymore!

Carbide 3D Shapeoko Pro and Inventables X-Carve Pro

The two most visible hobbyist CNC router makers look to be leveling up at the same time with new prosumer-grade devices. Expect beefier fittings along with beefier prices.

Creality CR-30 Belt Printer

After seeing a few people attempt to bring the "belt printer" concept to market, it looks like Creality will be releasing a consumer version soon. We've seen the White Knight (page 29), which shows lots of promise in this area, and you can find the files to build your own on Thingiverse (thingiverse.com/thing:3324280), but Creality has been putting out decent printers for a while now, so they should bring the manufacturing chops to make it happen. Knowing Creality, it will be pretty reasonably priced as a bonus.

For those unfamiliar, the design looks a little odd, but allows for a virtually unlimited single axis. The bed is a belt, which allows for you to move a print out of the way in order to automate printing multiple files, or to print something long — like a sword — that wouldn't fit on a static print bed.

Prusa XL

Now that the Prusa Mini (page 34) has launched, the Prague-based company tells us that they're going the opposite direction for their next machine, the Prusa XL. As the name implies, this will be a larger printer than the flagship Prusa i3 MK3S. Beyond that, we haven't learned much about any other potential changes (a different extruder nozzle perhaps?), but we're eager to see it materialize.

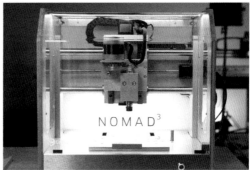

Carbide 3D Nomad 3

Several companies have put out desktop CNC mills this year, tiny units capable of doing light milling in aluminum, brass, and softer materials. We got a chance to play with the new Tormach and Bantam machines (see reviews on page 33), and now Carbide 3D has revealed, along with their Shapeoko Pro, an updated Nomad that should be available very shortly. Competition in this space is very exciting.

Rendyr Optic

Virginia-based startup Rendyr is gearing up to release its debut machine, a laser engraving and cutting device that offers fold-up, tuck-under-your-arm portability. It's a beautifully designed device with an 11.81"×17.13" cutting area, or you can remove the magnetic cutting bed to engrave directly on a flat surface of essentially any size. It also incorporates an internal three-stage air filter. The Optic's pre-release price of $2,000 is the cost of a Glowforge, but its 15W diode laser offers a fraction of a Glowforge's power (and it doesn't have that wonderful camera interface). We'll see if the portability is worth its premium. ⊘

CAD DESIGN TIPS FOR 3D PRINTING

Written and illustrated by Billie Ruben

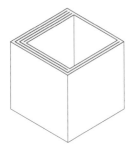

Make walls a multiple of your extrusion line width for a smooth slice. If it was 0.4mm use 0.8, 1.2, 1.6, etc.

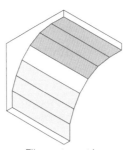

Filament must be laid upon existing material, so avoid steep overhangs to reduce the need for support.

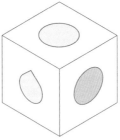

Vertical holes are fine, but horizontal ones should be tear-drop shaped to mitigate steep overhangs.

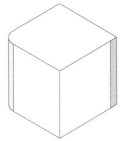

Vertical edge fillets increase quality by reducing inertia during harsh directional changes.

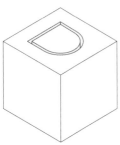

Roughly 0.3mm clearance should be added between fitted parts using offset face at end of modeling process.

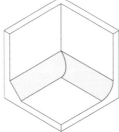

Adding a fillet or chamfer between a wall and base strengthens the join by adding more interface.

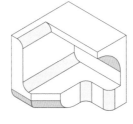

Fillets don't work well from below, due to harsh overhangs. But they can look great in other areas.

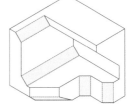

Equal chamfers always work (even from below) as their overhang remains at a printable 45°.

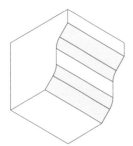

Combining fillets and chamfers mitigates the issues of fillets alone and smooths the chamfer.

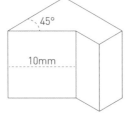

Using parameters and constraints allows you to easily edit and iterate upon your designs.

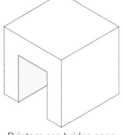

Printers can bridge gaps between bodies quite easily. Distance varies, but most can easily handle 2cm+.

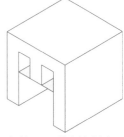

A thin, sacrificial bridging layer can reduces the need for support material. It is cut away after printing.

L like all production processes, 3D printing has constraints which are to be considered during the design phase in order to make high-quality, functional, and beautiful objects. I love seeing people make their own designs to fulfill their specific needs and desires, so I've collected many of these considerations to help you hit the ground running! ♥

BILLIE RUBEN is a life-long maker, ex-moderator of r/3Dprinting and current admin of the largest 3D printing Discord. Billie enjoys collecting skills and sharing them with others, particularly with 3D printing. You can support her at patreon.com/BillieRuben, or follow her on Twitter: @BillieRubenMake.

Sacrificial, perpendicular ribs can be added to support overhangs during printing.

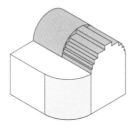

Curves look good with an axis in the Z direction, but due to the layering process can look very poor in the X/Y.

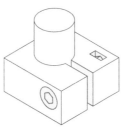

A slit, bolt, and trapped nut can be added to holes to allow them to be tightened around another part.

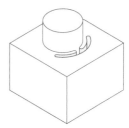

Compliance can be added to parts to enable flex, which enables push-fitting parts.

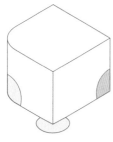

Reduce the risk of a print warping up from the bed by rounding out or adding mouse ears to corners.

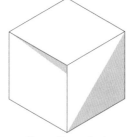

Use software that makes manifold objects (without tiny gaps or reversed faces), to avoid slicing errors.

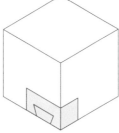

Complicated or fitted parts of an overall print can be isolated and printed to test for fit.

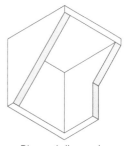

Diagonal ribs can be added to support/enable a roof to bridge between them. Can be beneficial inside a model.

Text looks best when indented into a vertical surface. It reduces overhangs and has better resolution.

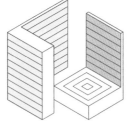

Due to the planar layering of most 3D printers, print orientation has a significant impact on strength.

This infographic is part of a series of 3D printing poster guides that I have produced. You can access full-resolution copies of these guides for free so you can print your own; I also sell prints through RedBubble if you'd like to support me making more. Links for both can be found here: billieruben.info/shop

If you get stuck with a design, the 3DPrinters community on Discord (discord.gg/B4tp8MH), which I help run, is happy to help you out.

Happy Printing! ✅

RESIN 3D PRINT SUPPORTS

CREATING SUPPORT STRUCTURES FOR YOUR SLA, DLP, OR LCD RESIN PRINTS — THE RIGHT WAY

Written by Dominik Laa
Translation by Niq Oltman

DOMINIK LAA studies business informatics at Technical University Vienna (TU Wien) in Austria. He's interested in additive manufacturing and the huge DIY community behind it. After getting his first 3D printer he started creating his own resin printers and other mechatronic projects.

Dominik Laa

When 3D printing in resin, you will need **support structures**, and it's hard to get them just right. Your slicer software can add them automatically, but this rarely works perfectly. Adding them manually requires quite a bit of knowledge. This guide will help you set up supports correctly and get to know the important parameters. We'll cover:

- Why you (almost always) need supports in resin prints, and where to place them
- How to assess auto-added supports, and how to improve them manually
- Differences between resin and FDM printing when setting up your prints

Even with the common FDM printing method using molten plastic, for many 3D prints to come out correctly, it's essential to add the right support structures. And if you're printing in liquid resin, you need to take even more care. Relying blindly on auto-added supports, or leaving them out entirely, is likely to cause problems.

Why are supports so important? In resin printing, a light source (laser, LED, or projector) is used to **cure** the liquid resin — to cause a chemical reaction that makes it solid. Like all 3D printing methods, this takes place layer by layer, but compared to FDM printers it's inverted. The current layer being cured is at the *bottom* of the resin container, which is transparent. When the layer is complete, the object being printed is moved slightly upward, peeling off the bottom of

the container, to make room for the next layer. Without adequate support structures to connect delicate parts of the workpiece to the (upside-down) printing platform, the freshly cured layer may get stuck to the bottom of the container and become separated from the workpiece. This will almost certainly ruin the print. Worse, it can damage the delicate film that forms the container bottom in many printers, causing critical leaks.

Choosing a Slicer

SLA printing typically uses **tree-type supports**, unlike the **lattice-type supports** often used in FDM printing. A piece of software called a **slicer** is used to prepare your 3D print job, including adding the necessary supports.

The features of some popular slicers for resin printing are compared in the table below. There are many more differences between these slicers. Their approaches to generating support structures vary greatly, as do the available options for tuning this process. Note also that none of the slicers shown is directly capable of pre-processing a 3D model for printing in resin.

In this guide, I've chosen PrusaSlicer as an example for demonstrating the important parameters. You can use this free software not only for the SL1 from Prusa Research, but for any other resin or FDM 3D printer.

For PrusaSlicer and all the others, you'll initially need to enter the specifications of your actual printer. Slicers do come with pre-set

FREE SLICERS FOR RESIN PRINTERS

SLICER	CHITUBOX	Z-SUITE	PRUSASLICER	NANODLP
Vendor	ChiTuBox	Zortrax	Prusa Research	NanoDLP
Platform	Windows, macOS, Linux	Windows, macOS	Windows, macOS, Linux	Windows, macOS, Linux, Raspberry Pi
Hollowing models	✓	✓	✓	✓
Automatic supports	✓	✓	✓	✓
Editing automatic supports	✓	✓	✓	
Adding supports manually	✓		✓	✓
Rotating models in 3D	✓	✓	✓	✓
Previewing layers	✓	✓	✓	✓
Controlling the print job				✓

profiles for some printers, but these usually include only the manufacturer's own products. For PrusaSlicer, this is the SL1; for Z-Suite, it includes Zortrax's resin printers.

Raft Layers

Almost all resin prints require support structures if you want them to come out correctly. For instance, look at the Benchy boat, a popular benchmark model for testing 3D printers. You'll find that it doesn't have any big overhangs; that's one reason it became popular for testing FDM printers. It's useful for testing resin printers as well — you usually don't need to worry about supports inside the model.

However, some support structures outside are still needed. PrusaSlicer generates them automatically if you slice the model using its standard settings (Figure **A**). When you're printing in resin, you'll almost always want to have a gap between the build plate and the object — 3mm to 10mm in height — to make room for a *raft* or *attachment layer* support (in PrusaSlicer it's called a *pad*.). That's because the initial layers of a 3D print in resin will be cured for much longer than the rest (by a factor of 3, even up to 10) to ensure that they're affixed very strongly to the build plate. Should your workpiece peel off the plate during printing, your print is inevitably ruined — just like FDM, only upside down.

As an added benefit, you're unlikely to damage your finished print when you take it off the build plate, which is usually done with a scalpel, utility knife, or metal scraper. The tool will only touch the supports, not the model itself. In PrusaSlicer the parameter for this gap be found under Print Settings → Support Material → Pad.

Overhangs

If you're printing an *overhang* with a slope of around 45° or more (depending on your material and your printer), you should definitely include supports. In the Benchy model, we can spot a nice example, which PrusaSlicer detected automatically (Figure **B**): The lip of the anchor hawse hole on the boat's sloped side wall forms a small overhang at an angle of roughly 90° — practically horizontal.

With FDM, this little overhang would be

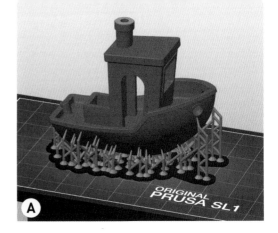

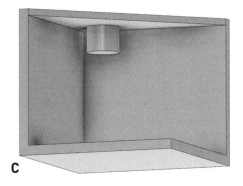

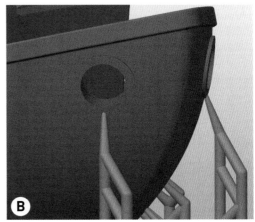

Dominik Laa

negligible, as the material is always suspended from the fresh filament being pushed out of the print head. But with resin printing, this protrusion may be too much.

There's much less of a problem where horizontal surfaces straddle a cavity with no supports in between, like a bridge. For example, the Benchy cabin roof can be printed without adding supports because it's not very long.

Islands

When models are sliced, some layers may contain parts that aren't connected to anything else in the layer, or to the building platform (Figure **C**). Without supports, these *islands* are doomed to fail; they'll remain stuck to the bottom of the container. You can detect islands by inspecting each layer in the slicer, looking for parts that hover freely in the air (Figure **D**). However, you can typically avoid this scenario entirely by adding automatic supports.

Another critical situation is where large areas are connected to the rest of the object in just a few places. When a part like this peels off the resin tank, the adhesive forces are particularly strong, so you should add additional supports to make sure these areas are bonded well to the rest of the model.

Loading and Orienting Models

Once you're done configuring your slicer to work with your printer, loading a 3D model works just like you're used to with FDM printers. The next step is important, though (and many users don't do it): for resin prints, unlike FDM, you should **rotate** a flat-bottomed model so that its underside is slightly tilted — not parallel to the printing platform. If you leave the model as-is after importing (Figure **E**) and don't change the orientation, the forces caused by the initial flat layer (Figure **F**) detaching from the bottom of the resin tank may deform or even rip apart your model — despite having proper supports.

Rotated slightly, a model's large base surface is less of an issue, as there is less surface area per layer that needs to peel off after being printed (Figure **G**). In addition, you'll need far fewer supports. The layers in the final printed model won't be horizontal, but as resin printing typically creates

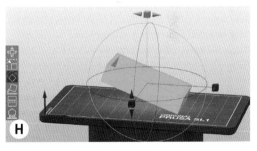

very thin layers, you're unlikely to notice this.

For most cases, I recommend rotation by 10° to 30° (rotating around a single axis is sufficient). To do this in PrusaSlicer, press the R key or click on the Rotate icon at the left edge of the screen (Figure **H**). PrusaSlicer can also apply rotation automatically (right-click on the object, and select Optimize Orientation). However, this tends to rotate the model in unexpected ways, which may greatly increase the time for printing as well.

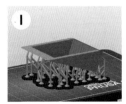

GOOD: Tilt the box for smaller cross-section areas.

BAD: Strong peeling forces on large area parallel to print bed.

GOOD: Support large free-hanging areas (box top hidden in this view).

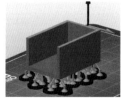

BAD: Printing large free-hanging area without supports.

GOOD: Abundantly support over-hanging parts in exposed areas ...

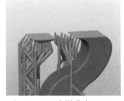

... to secure any initially loose "islands" in the model.

BAD: Supporting overhangs sparsely or negligently leads to ...

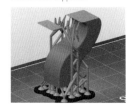

... loose islands that will stick to the tank bottom during printing.

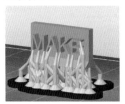

GOOD: Be sure to support large horizontal overhangs.

GOOD: Tiny horizontal overhangs do not need to be supported.

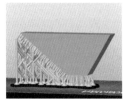

GOOD: Support long overhangs, even around 45 degrees.

BAD: Don't try to support long overhangs with a small footprint.

Dominik Laa

Generating Supports

In PrusaSlicer, it's super easy to get automatic support structures. There's a Supports setting in the right-hand panel to choose if you want supports only between the build plate and the model ("Support on build plate only"), or between different points in the model as well ("Everywhere"). The first option is often sufficient for simple models such as the Benchy boat.

Click on Slice Now, and the software will generate the appropriate structures. Until you slice, the model will be shown resting on the build plate; after slicing, you'll see that the raft layers mentioned earlier have now been applied, too.

The auto-generated supports usually aren't bad, so it's a good idea to create automatic supports first, and then add or remove supports manually as needed. You should do this wherever your previous prints had problems, or where you find some of the typical issues we mentioned above: flat or long overhangs, large surfaces, or loose parts (islands). Figure **I** gives an overview of some things that have turned out to work well (Good), and some that didn't (Bad). Supports are

shown in grey and the workpiece in green. Some objects are only partly shown to make things clearer — try to visualize them extending past the orange-colored cut surface.

You can click on the model to add or remove supports, and you can change global settings for the supports in Print Settings → Supports. Figure **J** shows you the components that make up a support structure. Every support structure has a **base**. It needs to be large enough to remain stable under the weight of the remaining structure (2–5mm in diameter and 0.5–2mm in height, depending on your material).

The support **pillar** connects the base to the head. You can set the pillar's diameter. Advanced slicers can also **link** or **join** several pillars automatically to create a more stable structure; in PrusaSlicer you can set a maximum "bridge length" and "pillar linking distance," for example.

The part that connects the support and the model is known as the **head**. To ensure a secure connection, it will typically penetrate 0.1 to 0.7mm *into* the object. The head diameter at the top (normally between 0.2 and 1mm) controls the

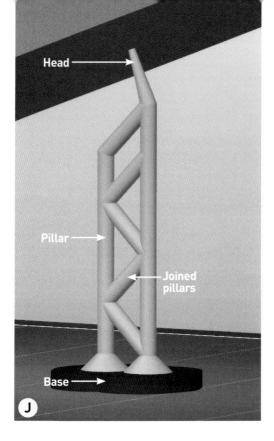

Head

Pillar

Joined pillars

Base

J

amount of force that the support structure can take. Be aware that large structures will tend to leave imperfections on the printed object. Confusingly, the length of the head is labeled as the head width. You shouldn't make it too long — 2mm to 5mm is known to work well.

In general, you can depend on PrusaSlicer's standard settings for these support parameters; then if there's a problem, you can work your way toward the right values. ⊘

For more tips on resin print hollowing, slicing, exporting, and printing, read the extended version of this article at makezine.com/go/skill-builder-resin-3d-print-supports.

KEY TO GOOD PRINTS:
YOUR PRINTING ENVIRONMENT

Proper supports are critical for resin printing, but it's also important to have the right environment for printing. For one, your resin should not be too old — when stored correctly (as per the instructions on the bottle), most resins should last you for about a year. Before filling your printer, give the resin bottle a thorough shake so the included pigments can distribute evenly. After shaking, let the bottle rest for a while to allow air bubbles to escape.

Keeping track of temperature is essential, too, and often under-appreciated. In my experience, printing in a cold environment such as your basement workshop or an air-conditioned server room tends to produce many failed prints. It's much better to print where it's warm — preferably around 77°F (25°C) or slightly above. Better still is to have integrated heating elements that warm up the air inside the 3D printer, or warm the resin directly. Depending on your material, the air should reach a temperature of 95°–122°F (35°–50°C); for the resin, 85°–105°F (30°–40°C). Until you've got sufficient experience, be careful fooling with heat settings. While a print is in process, the resin and ambient air will usually keep a stable temperature, as the illuminating display will create sufficient heat all by itself.

RECYCLE YOUR REJECTS

Shred those failed 3D prints and re-extrude the plastic to save waste and money

Written by Samer Najia

Five tries. That's how many times on average I prototype something before I have a workable 3D printed design. And that doesn't include failed prints, running out of filament at the wrong time, or printer defects such as warping. I'm no engineering prodigy: I make tons of mistakes and I make them often. That means I end up with a lot of waste, not just failed prints, but also support material, brims, rafts, and parts that break and have to be reprinted.

Because of all this waste, I like to print with PLA because at least that will biodegrade (eventually). These days, though, that's not enough. Based on weight alone, I figure about 1 reel of filament out of every 6 is lost as scrap. That's material that ends up in a landfill or incinerator, which seems like a colossal waste.

So, what can you do with your 3D printer waste? You can:

- Throw it in your home recycle bin and let whoever does your curbside pickup "handle it" (they can't)
- Take it to a recycling center and ask if they can do something useful with it (probably not), or
- Recycle it yourself — by shredding and using it again and again. The U.S. Army's doing it, the International Space Station's doing it, and you can too.

Recycling Printer Plastics

Let's take a deeper dive with recycling your 3D printer waste. I print mostly with PLA but I still occasionally use ABS and flexible materials like SemiFlex and NinjaFlex. I also have aspirations to print with PETG and other materials. Spools may be made from HIPS or ABS, and those also have potential to be turned into filament material. Aside from PLA, none of these plastics will biodegrade — recycling is the way to go.

Shredding

Recycling your failed prints starts with shredding. The material has to be reduced to pieces that can be fed reasonably into some sort of hopper, which means no more than 5mm in diameter. This can be done in a few ways:

- Smash up your waste plastic with a heavy hammer before putting it through a sturdy paper shredder or heavy duty kitchen blender.

My multiple attempts to create a carriage for the Lily printer head.

Samer Najia, Flowalistik

The open-design Precious Plastic shredder used by Agustin "Flowalistik" Arroyo as part of his Plastic Smoothie recycling initiative in 2020. #plasticsmoothie

There are a number of Instructables and other guides on the internet, some of which take you through modifying an ordinary paper shredder for small volume work.

- Build or buy a heavy duty shredder like the one shared by Precious Plastic (preciousplastic. com), an open source project that's worthwhile if you generate enough plastic waste. Others include the Filabot Reclaimer or 3devo Shr3dIt.
- Find a local group or makerspace with a shredder and/or granulator. I located a group within a local university's engineering program who were willing to do the shredding for me (thanks University of Maryland!).

SAMER NAJIA holds a degree in mechanical engineering from Duke University but making things is his true passion. He spends countless hours building progressively larger and more complex projects. He's co-author of the new book *Mechanical Engineering for Makers: A Hands-on Guide to Designing and Making Physical Things*, available at makershed.com/books.

Bags of granulated plastic shred.

Mixing With New Material

Once you have your shred material you can mix it with pellets of virgin material (i.e. never been used as filament) to create a formulation to your liking. You can add dyes of various kinds to change the color (but these can also impact the material properties) or you can throw in add-ins like super-fine sawdust or metal powders (both of which are abrasive enough to eventually wear down your brass nozzle).

I like to experiment with various mixes, in both PLA and ABS, to see if I can improve the filament's ability to resist humidity, or add strength or resistance to heat. For instance, I have noted that certain colors are more prone to extruder clogging, certain additives adhere better to the print bed, and "wood"-infused PLA stands up to direct sunlight much better than plain PLA.

Commercial recycled filament (see page 50) can contain up to 50%–70% recycled PLA. But even if your mix ratios might be 80:20 virgin to ground material, reusing your plastic waste like this reduces filament costs, and of course reduces environmental impact.

Why mix with unmelted material? If you print with recycled PLA exclusively, your prints may be more susceptible to warping and won't have the same tensile strength. It's easiest to add more virgin material. However, warpage can also be controlled by designing smaller surface areas and using brims, rafts, and supports, and strength can be improved by painting the finished part with resins.

One great use case for recycled PLA is to make casting masters for "lost PLA" metal casting. The PLA is going to be vaporized in the mold creation process anyway, so strength is not a factor.

While PLA is more benign environmentally than ABS (ABS doesn't biodegrade at all), it can still take centuries to degrade in a landfill. We as makers may be contributing only a small amount to the mountains of plastic waste around the world, but we should do our part to reduce and recycle as much as possible.

Recycling Paths: Filament or Pellets?

So let's melt some shreds! For the most part, there are two basic recycling paths for your shredded prints:

1. Make your own filament. Pellets and waste material are fed to a machine with an auger that melts and extrudes the plastic into new filament that is collected on spools. While the filament properties may be different, the operational aspects of your printers are unchanged. Note that with this method, plastic is melted *twice* post grinding: once to extrude it into filament and again when being consumed by the printer.

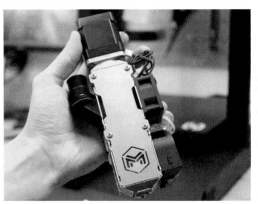

The Filabot EX2 filament extruder is spendy but reliable. You can pick up refurbs for less than $1,800. filabot.com

The Mahor XYZ pellet extruder, about to be deployed in Flowalistik's custom Plastic Smoothie 3D printer. mahor.xyz

Samer Najia, Filabot, Flowalistik

2. Extrude pellets on demand. Your printer's extruder consumes plastic waste and pellets instead of filament, and extrudes the melt directly on the build surface. Compared to other printers, the pellet print head is typically heavier, slower, and bigger, leaving a smaller print volume.

I have worked with two recycling systems that cover each use case: the Filastruder filament extruder (Figure **A**) and the Lily direct pellet extruder (Figure **B**). The Filastruder is available from filastruder.com as a kit, can be mounted horizontally or vertically (it's better vertically), and can be assembled to include a takeup spool that matches the speed at which filament is exiting the Filastruder. There are similar systems, such as the open source Recyclebot, the heavy duty Filabot, and other commercial products. The idea remains the same — the end product is filament that can be used in multiple printers.

The Lily Pellet Extruder from Recycl3dprint (recycl3dprint.com) involves modifying a printer to directly consume pellets — no spools, no additional systems. The extruder and its pellet dispenser are a complete platform to directly print objects. The Lily also lets you swap nozzles like other printers, but without worrying about filament adjustment.

Both systems have their benefits and drawbacks; which one fits for you depends on what your needs are. In my case, I needed both. With the Lily, I wanted a printer that can handle scrap material immediately so that I can experiment with filament on the fly (changing the material composition on demand), and I wanted to add 3D printing capability to my Ox CNC and possibly my Shapeoko; both have very large X-Y work surfaces, run slowly (10–30mm/s), and are heavy duty enough to take on a heavy Lily head. I also wanted to experiment with blends that include ABS, leftover resin, and possibly composites like fiberglass.

My Filastruder's primary purpose is to make filament to experiment with. I typically fabricate specific blends in bulk and then test the filament on multiple printers. If I were to build mine again, I would create a larger enclosure than the kit one, to allow easier access for my stubby fingers to get to the wiring. I also purchased the winder, which includes sensors to control the speed at which

A

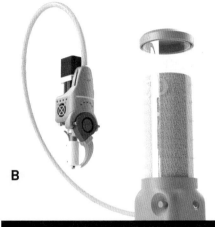

B

RECYCLED PROSTHETICS

Enable Alliance is doing great things with recycling. From creating e-NABLE prosthetic hands using renewable energy, to turning e-waste into 3D printers and scanners, their "Project Alchemy" and ReFab Dar (refabdar.org) have helped many people all over the world with research, assistive devices, and recycling technologies. For more information about all these technologies, or to participate in their efforts, contact info@enablealliance.org.

BUYING RECYCLED FILAMENT

Not ready to grind and re-melt your own plastic junk? A handful of vendors sell recycled filament in a variety of popular materials. Some of these are made from factory waste, and some are actually post-consumer. Here are all the options we could find in 2020.

Vendor	PLA	ABS	PET	PETG	HIPS	Notes
Filamentive filamentive.com	50%–70%	64%	100%	67%–100%		rPLA & rPETG from factory waste; ONE PET 100% post-consumer (suspended for Covid-19); also 100% rPET/carbon fiber and 50% rPLA/wood
Filabot filabot.com				100%		From polypropylene/polyester fabric
RePLAy 3D replay-3d.com	100%	100%				From factory waste
Closed Loop Plastics closedloopplastics.com					100%	Pink only, from post-consumer red party cups
Tridea tridea.be/shop	100% pre-order		100%			ONE PET 100% post-consumer (suspended for Covid-19)
3D-Fuel 3dfuel.com	100% pre-order		100%			rPLA from factory waste; also virgin PLA with recycled bio-fillers like hemp fiber, beer grains, coffee grounds

RePLAy 3D

the spool takes up filament, although I haven't yet built it. I tend to let the filament just fall to the floor and roll it up later.

With either type of system, you can experiment with recycling not just 3D prints but other printable plastics, like PETG water bottles.

I used my evaluation unit of the Lily as an excuse to finally build my SmartAlu printer (thingiverse.com/thing:1846437). The design was easily adapted to use the Lily: I replaced the V-wheels with linear rails, turned the motor and pulley mounts inward inside the case, and fabricated a carriage mount for the Lily. Don't let that scare you off, it's not a big deal, especially if you're a tinkerer. While my mount is made of PLA, you can also fabricate it out of metal. You just have to work out a finite number of design details. In the end, I lost 100mm in each axis because of the Lily's size, but I have it on good authority that a slimmed-down version is in the works (and I could've kept my 300mm×300mm×300mm print volume by using longer extrusions). If you like to tinker and be creative, this is the kit for you.

I recommend that you have two printers if you're converting one, or that you build one specifically for this. The Lily has been tested on a number of popular printers, such as the Ender 3 — contact the vendor at Recycl3DPrint.com for details of implementation. The vendor will also work with you to make minor changes to your firmware, so be sure to have a backup of your existing firmware if you plan to roll back your changes.

I'm still testing the Lily — check online (makezine.com/go/lily-pellet-extruder) for updates and video. If time permits, I'll also mount a Lily on one of my CNC machines and see how it behaves on a machine that uses leadscrews, not belts.

Of course, plastic isn't all I recycle — in the next issue we'll explore my 3D printers built from recycled parts. ⊘

READ MORE:
"Closing the Loop on 3D Printing" from UC-San Francisco Maker's Lab: library.ucsf.edu/news/closing-the-loop-on-3d-printing

3DP Recycling overviews: all3dp.com/2/3d-printer-recycled-plastic-tips-for-your-waste-plastic and all3dp.com/2/the-3d-printer-filament-recycler-s-guide

CONVERT A PRINTER TO PELLET EXTRUSION

10 top tips for making the switch Written by Samer Najia

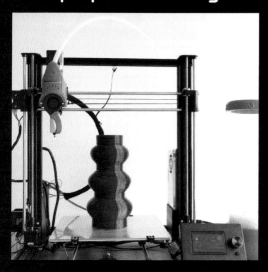

Converting a 3D printer to print directly from pellets requires tradeoffs: you'll gain the ability to reuse plastic material in exchange for some performance limitations. Done carefully, these modifications will open the door to closed-loop recycling, economical pellet feedstock ($2–$5/kg), custom material blends, colors, and properties — but they could reduce the usable print volume and require slower print speeds due to the toolhead's weight. Ten things to consider:

1. Can the carriage handle the weight and dimensions of the pellet extruder and hot end? Consult your vendor(s). We mounted the Lily pellet extruder on a SmartAlu printer designed by Smartfriendz (above).

2. Obtain a copy of the firmware on your printer — you may need to modify it. A lot of modifications can be stored in your slicer profile, but various offsets may have to be changed in firmware.

3. Your Z-axis endstop may have to move if the pellet extruder has a long hot end (the Lily does).

4. If you use a BLTouch or induction sensor for bed leveling, you might have to redesign a new X-carriage to accommodate these elements.

5. If possible, make the conversion on a secondary printer in case you need your main printer to prototype parts.

6. Decide how you will feed the pellet extruder. If you're using a gravity feeder, the hopper has to be above the printer. The Lily pushes filament through a PTFE tube using air, so the pellets or shreds can be fed from any level.

7. While V-wheels are great, expect wear to increase with the heavier head. Consider upgrading at least the X-axis to use a linear rail.

8. If your kit comes with a MOSFET, use it. Running the pellet extruder may consume more power than your printer's original design. Make sure your power supply can support the additional load.

9. Once it's all installed, you'll want to experiment with different materials to determine optimal print speeds, jerk settings, and retraction rates.

10. Expect to tune your PID values to suit your extruder. Recycl3D provided me with a baseline set of numbers and I tuned accordingly.

Whatever extruder you choose, seek support from your vendor, as this is a non-trivial endeavor. They'll have complete details — Recycl3D was in close contact throughout my Lily installation over several weeks. Of course, if you're using a DIY pellet extruder, the same checklist applies, but otherwise you're on your own. You can find many #pelletextruder sources on YouTube, Thingiverse, and other sites. ✏

PERFECTING PORCELAIN PRINTS

Written by Tom Lauerman

CLAY-BASED PRINTERS CAN MAKE MUCH MORE THAN VASES AND POTTERY. HERE'S HOW TO GET STARTED

Cody Goddard eli.aanda.psu.edu/cody-goddard, Tom Lauerman

TOM LAUERMAN
is an artist and educator
based in Pennsylvania.
tomlauerman.com

Many folks think primarily of pottery forms when they think of 3D printing in clay, but in fact one can print a very wide range of object types including architectural components, human figures, and abstract sculptural forms. Here are some aspects to consider when getting started digitally fabricating with this medium.

How Clay 3D Printers Work

There are two ways of pushing clay through an extruder. One approach, which is cheaper and requires a bit more analog "finesse," uses compressed air to move the clay along. Generally, these systems use a somewhat softer clay as it is a bit easier to move. These are often coupled with delta-format printers, as the non-moving build plate is an advantage with softer material. The second approach uses a mechanical feed via a very powerful motor moving a plunger through a tube (Figure **A**). This tends to be more expensive and adds some engineering complexity to the design, but can allow for the use of stiffer clay and the extrusion is controlled via software as it would be in plastic printing. I prefer a mechanical system and find it allows more fine-grained control in my printing. However, an attentive and careful user can get great results from either system, just as poor results are possible with either.

Most people printing clay use nozzles that are 1mm or larger, sometimes much larger — 3mm–4mm, and well beyond (Figure **B**). Everything happens at room temperature, so one never has to worry about a melt rate slowing things down or material "cooking" in the nozzle. There are also no fumes, emissions, or impacts on air quality from partial combustion. With a large nozzle, like 3mm, and a spiral form that is one continuous perimeter, the printing can go exceptionally fast relative to plastic printing, with a cup or vase materializing in something like 10 to 30 minutes (Figure **C**). However, finer nozzles, infill, and multi-material approaches all slow things down toward plastic-printing timeframes.

Creating Models

To design for printing with clay, the familiar software tools are often used, including Blender, Fusion 360, and Rhino3D. There are a number

of folks who like to use Rhino + Grasshopper to generate spiral toolpaths directly as gcode output. I use whatever program is best suited to the forms I am making, with my favorites being Blender and Rhino. I don't design differently for clay, but I am a bit unusual in this regard as I'm personally a little less interested in hollow "spiral vase" forms and more interested in architectural, sculptural, and geometric abstract shapes (Figure **D**).

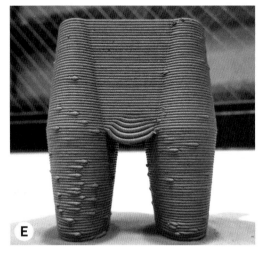

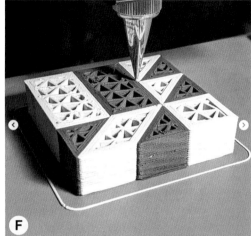

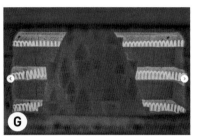

Tom Lauerman, Cody Goddard eli.aanda.psu.edu/cody-goddard

For slicing, clay printing people use the same programs one sees in plastic FDM. I use Prusa Slicer as I also have a few Prusa FDM printers and have been impressed by the program's evolution and enjoy that it is open source.

I often "preview" forms in plastic FDM before moving to clay. A problem in plastic is likely to be a problem in clay too, and often more severe. Overhangs can be quite similar to plastic, in terms of angles that can be attained. However, bridging is more effective in plastic than clay (Figure **E**), as plastic cools so quickly and becomes rigid.

One significant difference between plastic and clay printing is that clay shrinks a lot as it dries and again as it is fired. This can range from something like 5% shrinkage all the way to something approaching 20% shrinkage with certain types of porcelain clay.

When working with clay, the material can take many hours to dry. This is an advantage in my view — one can go in with fingers or with tools and change the form, cut, join, paste, bend, smooth, or perform any number of other "post-processing" manipulations.

As for internal mass, I have not found any particular restrictions. I use infill as I would in plastic (Figure **F**), and this is able to dry out and go through the kiln firing process just fine (Figure **G**). Clay is a great deal heavier than plastic, more dense. So the objects one prints have a real heft, as one would expect from ceramic objects like pottery forms, sculpture, or bricks. However, one could make a very lightweight and highly insulative brick by using infill. Presumably this could still be quite strong, like a traditional building component.

Printing and Processing

After designing and testing my model in plastic, I choose what type of clay I might work with, the most common types being earthenware, stoneware, and porcelain. Each has advantages, and all are a great deal cheaper than plastics. I sometimes dig clay in my backyard for printing material (Figure **H**). I prepare the clay to just the right consistency by adding water or wedging the clay on a plaster block to remove water.

When ready, I load the clay into a polycarbonate tube which I then place in the printer, and begin printing. I make sure to soak the build plate in water so the clay has a damp surface to adhere to and won't pop loose as it dries. I use clay build plates presently, but have used plaster or wood in the past (Figure I). It is helpful to use a very porous material for the build plate to aid in the drying process. When printing on glass, plastic, or metal, the clay touching the build surface will dry extremely slowly or not at all.

After printing I remove the build plate and cover the object with a small plastic bucket, glass bell jar, or in a pinch, just a plastic bag to keep it from drying too fast. Clay can crack and warp badly if dried quickly and/or unevenly. I dry things slowly. I like to just forget about them for a few days and move on to the next print. So a controlled drying time might be as little as one day but is often weeks or longer. When the clay object is bone dry (also known as "greenware") it is ready to be placed carefully in a kiln and fired. If it is intended to have a glazed surface it will typically be fired twice; if it will remain unglazed only one firing is necessary. Clays can be fired in a range from about 1,800°F to over 2,300°F, and the resulting objects are permanent ceramic and will likely be with us for thousands of years, hopefully in someone's collection. ⊘

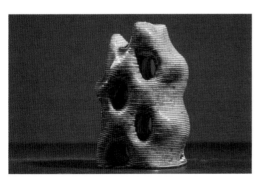

DOUBLE TROUBLE

I have experimented a great deal with a dual extrusion system, using two colors of clay in two print heads. This requires a lot of calibration but has worked effectively and has a lot of future potential in my opinion (Figures J and K).

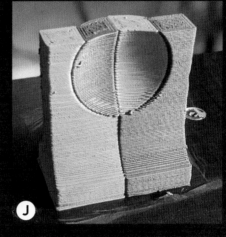

J

K

Kodi Smart TV Streamer

Tired of regular old TV? Turn a Raspberry Pi into a media streamer for movies, music, Netflix, Prime Video, and more

Written by Heinz Behling • Translation by Niq Oltman

Now that we're all forced to spend more time at home thanks to the pandemic, we're seeing the limits of what regular TV can do for us in terms of entertainment. *The Mummy*, again, really? Even binge-watching all 279 episodes of *The Big Bang Theory* will eventually get boring. It's time for something new. Popular streaming services like Netflix and Amazon Prime Video promise an easy cure.

You don't need a smart TV, all you need is a Raspberry Pi 4, a bit of software, a not-too-shabby internet connection (25Mbps for Ultra HD), and a subscription to a streaming service, and you're set to enjoy high-definition video entertainment, even up to 4K UHD. This guide will show you how to get everything up and running, so you won't have to watch another rerun.

MAKE ANY TV SMART WITH FREE SOFTWARE

For the media player software, we'll be using Kodi, version 18 (Leia) (Figure Ⓐ). Kodi is free, open source home theater software that plays virtually all formats of video and music, but it can do much more. Using add-ons, you can greatly extend its features, stream internet content, and even use it as a platform for gaming (but that's outside the scope of this guide).

To run Kodi we'll install LibreELEC, a lightweight Linux OS distribution built specifically

TIME REQUIRED:
1–2 Hours
DIFFICULTY:
Easy/Intermediate
COST:
$70–$80 (plus subscription costs)

MATERIALS
» **Raspberry Pi 4 single-board computer** with internet access and at least 2GB of RAM; 4GB recommended
» **Heatsink or cooler for Raspberry Pi 4** such as the Flirc Raspberry Pi Case, Amazon #B07WG4DW52
» **MicroSD memory card, at least 32GB** 64GB recommended
» **USB-C power adapter** for Raspberry Pi 4
» **HDMI cable with micro HDMI plug/adapter**
» **TV with HDMI jack and CEC-compatible remote** (see page 59)
» **Netflix and/or Amazon Prime Video subscription** first month free, then $9/month for Prime Video, $9–$16/month for Netflix

TOOLS
» **Computer with internet access** to set up your Pi; computer not needed afterward

HEINZ BEHLING is an editor for *Make:* Germany in Hannover who's into 3D printing and laser cutting. He started out long ago on the Commodore VC20 and C64; today he's 61 but still not an adult.

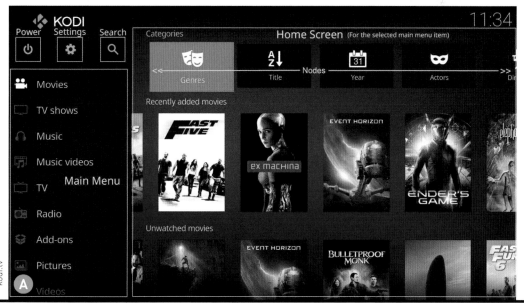

kodi.tv

Ⓐ

for Kodi, on a Raspberry Pi 4 single-board computer. Then we'll show you how to configure Kodi to allow add-ons from "unknown sources," and install the Netflix and Prime add-ons.

There are different approaches to installing Kodi on the Raspberry Pi: you can add it as a software package to your existing Raspbian system, or you can use a custom OS distribution called LibreELEC, which is built around Kodi. This latter method offers some benefits — it already comes with a number of useful software components that enable, for instance, using it with your TV's remote, and playing hi-res video. That's why we're using LibreELEC for this guide.

SET UP YOUR KODI STREAMER ON RASPBERRY PI

1. INSTALL LIBREELEC

It's really straightforward to install LibreELEC on a memory card. A tool called LibreELEC USB-SD Creator will handle all the required steps for you.

Simply download the tool onto your Windows, Mac, or Linux PC from libreelec.tv/downloads_new. Run it using administrator privileges, then select the latest LibreELEC version for Raspberry Pi 4 (version 9.2.5 "Leia" at the time of this writing). Use it to download the LibreELEC image onto your PC, then select your SD card, and write the image onto the card (Figure **B**).

A note about that memory card: In time, you'll have a sizable collection of music and video files on your Pi-based media machine, so you'll want to get a card with ample space. But if your card is more than 32 GB, there's an issue you'll have to work around: as required by the SD card standard, such cards will use the exFAT file system. Unfortunately, the Raspberry Pi's bootloader can't use exFAT; it can only run from FAT16- or FAT32-formatted media. Please refer to raspberrypi.org/documentation/installation/sdxc_formatting.md for a guide on getting your RasPi to boot from cards larger than 32 GB.

2. CONFIGURE FOR 4K (OPTIONAL)

After finishing the installation, don't remove the memory card from your card reader just yet. To make the Raspberry Pi work with 4K displays, you'll need to add a line of text to a configuration file on the card. Use your PC to open the partition named *LIBREELEC* on the card and look for the file called *config.txt*. Open it up in a text editor and add the following line to the end of the file:
`hdmi_enable_4kp60=1`

Don't forget to save the file, and to unmount the card from the system before you remove it.

3. CONNECT THE HARDWARE

Make sure your Raspberry Pi is powered off — the power cable should be unplugged — and insert the microSD card into the Pi.

Connect the Pi to your TV's HDMI input. Your HDMI cable should come with a micro HDMI plug on one side for plugging into the Pi; if not, you'll need an adapter. On the Pi, plug the cable into the port labeled HDMI0 — only this port supports the CEC features that allow you to control Kodi from your TV's remote (see "What's CEC?" to the right).

Not all TVs support CEC. Check your manual or on-screen menu; you might need to set an option to turn on CEC. Some TVs have a setting that allows waking up from standby as soon as there is a signal on the HDMI input. Take some time to set up your TV according to your preferences.

With your TV turned on and waiting for a signal on its HDMI input, now plug in the power cable on your Raspberry Pi. You should see Kodi's startup screen after a few moments. If your TV supports CEC, you can make all the requested settings

What's CEC?

CEC is short for Consumer Electronics Control. It allows you to control up to 15 compatible devices from a single remote control, typically the one for your TV. Depending on the brand of product, CEC comes under different guises: Sony calls it Bravia Sync, Philips uses EasyLink, and Samsung's name for it is Anynet+.

Once CEC has been enabled in your TV's menu, the signal from your remote is passed on to all other devices via the HDMI cables. This way, you can control your DVD player, for instance, without having to use a different remote. The transfer of control signals also works in the opposite direction: some multi-media players or sound systems can wake up your TV automatically as soon as you turn them on.

To use CEC with your CEC-enabled devices, you need to link them with HDMI cables where pin 13 is internally connected between the plugs. This is most likely the case for all cables made today. If you can't get CEC to work using an ancient HDMI cable, it's likely missing the pin 13 wire.

using the arrow keys and OK button on your TV's remote. If you don't have CEC, you can use a mouse and keyboard connected to your Pi via USB, or a wireless keyboard with a touch pad.

4. NETWORK SETTINGS

On setting up Kodi, you'll need to choose a network connection. Cable and Wi-Fi networks are supported. If you're using Wi-Fi, you'll need to pick your network name from the list and then enter your Wi-Fi access password, so make sure you have this information ready.

You'll probably want to be able to communicate with your Raspberry Pi via your network, such as for moving or copying files. This is enabled by turning on access via SSH and Samba in the

C Setting up network access via Samba server.

D Enabling SSH and setting a new password (to prevent unauthorized access to your Pi).

System screen, which you'll find by clicking on the gears symbol located on the left above Kodi's main menu (the media column). Select LibreELEC → Services → Enable Samba (Figure C). We highly recommend that you also turn on Samba password authentication. This will show you the current (default) Samba username and password. It's advisable that you change at least the password now. Write down the username and password that you chose.

Scrolling down a bit in the same screen, you'll get to the SSH settings (Figure D). Again, enable this service, then enter a new password (the default is *libreelec* for user *root*).

Change to Connections in the left column. On the right-hand side, you'll see the active network connection with its IP address. Make a note of it, as you'll need it later on for accessing Samba and SSH. You'll need Samba access for installing the Amazon Prime Video add-on. Configure your home network router so that it always assigns the same IP address to your Raspberry Pi (which will appear in the network as "LibreELEC").

5. CONFIGURE KODI ADD-ONS

Exit the LibreELEC section (using your remote's Exit or Back button or your backspace key),

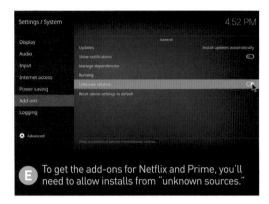

E. To get the add-ons for Netflix and Prime, you'll need to allow installs from "unknown sources."

F. This add-on is required for decoding and playing DRM-protected videos from Netflix and Amazon.

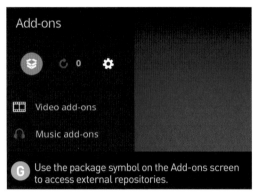

G. Use the package symbol on the Add-ons screen to access external repositories.

H. These add-ons will be installed in addition to Netflix.

then select System and then Add-ons. Click on Unknown Sources (Figure E).

Exit the System section and select Add-ons → Install from repository → LibreELEC add-ons → Videoplayer InputStream Add-ons. Click on InputStream Adaptive, then Install (Figure F). Choose the latest version for the add-on. Now go back to the System screen.

SET UP YOUR FAVORITE STREAMING SERVICES

6. SUBSCRIBE TO STREAMING SERVICES

Now it's time to subscribe to Netflix (netflix.com) and/or Prime Video (amazon.com/gp/video/storefront) if you haven't already. Note that these subscriptions are only free for the first month; after that, there's a monthly fee of $9 or more.

7. INSTALL THE NETFLIX ADD-ON

First, we'll install the repository for the Netflix add-on in Kodi — so that Kodi knows where to get the add-on. In the System screen, launch the file manager and choose Add Source. Click on <None>, then enter the following address:

https://castagnait.github.com/repository.castagnait/

Using the OK or Back buttons on your remote, move back to Kodi's main screen and select Add-ons. In the add-ons screen, click on the package symbol in the upper left corner (Figure G), then select Install from ZIP File. Pick the CastagnaIT repository, then choose the file named *repository.castagnait-1.0.1.zip*. After some time, a notification should momentarily appear in the upper right of the screen, informing you that the repository has finished installing.

Now you can Install from Repository: choose CastagnaIT Repository for Kodi 18.x (Leia). In the next screen, you'll find the Netflix add-on under Video Add-ons. Click on it for installation, picking the latest version if there are multiple choices. Once done, you'll be presented with a list of additional add-ons that are required to complete this installation (Figure H). Click OK to confirm. Depending on your internet speed, this installation may take a while. When finished, you'll get a confirmation screen (Figure I).

To sign in for the first time, click on the Netflix

I Congratulations, your Netflix install is complete.

J Playing copy-protected Netflix videos in Kodi requires a software decryption module called Widevine (from Google). It's extracted — slowly — from a ChromeOS package that's 3GB in size.

IMPORTANT: Netflix videos are decoded in software, requiring lots of computing power. This will make your Raspberry Pi's CPU break quite a sweat. A CPU cooler is highly recommended. If you haven't installed one yet, you should do so soon.

item and choose Open on the next screen. Enter your email address, click OK, then enter your Netflix password (case-sensitive). You'll now be signed in to Netflix, which will complete with a confirmation screen.

Click on your Netflix username to get into the Netflix menu. You've arrived! Pick one of the videos shown and launch it via (double) click. The first time you do this, you'll get a notice that this requires the Widevine content decryption module or CDM (Figure **J**). Click on Install Widevine, then confirm the licensing terms by clicking "I accept." (Don't be surprised that this is a Google license — Widevine is a Google product, taken here from Google ChromeOS recovery files.)

You'll get a warning that installing Widevine requires 3.1GB of storage space (remember our advice about getting a big memory card?). Confirm, and prepare for a long wait. Even on a hefty 150Mbps connection, the download will take about 3 minutes (Figure **K**). Unpacking the completed download will take another couple minutes.

Once everything is unpacked and installed, reboot your Pi. You'll then find Netflix under Add-ons in the media list (Figure **L**), ready for your browsing and viewing pleasure!

L Chill time: You've got Netflix now!

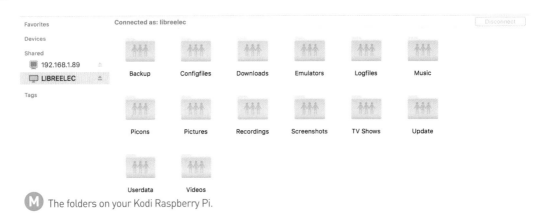

The folders on your Kodi Raspberry Pi.

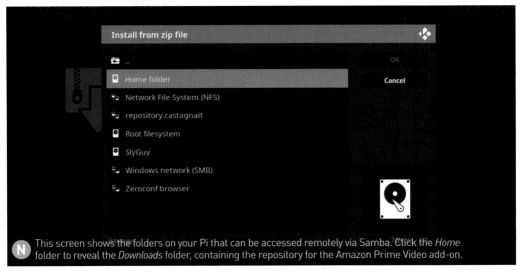

This screen shows the folders on your Pi that can be accessed remotely via Samba. Click the *Home* folder to reveal the *Downloads* folder, containing the repository for the Amazon Prime Video add-on.

NOTE: Like many Kodi add-ons, the Netflix add-on is something of a moving target; when the streaming company makes changes to its website, this Kodi add-on may need to be changed to keep up. If you are unable to log in using your correct Netflix username and password, there's a workaround: you can generate a Netflix authentication key. (That's the great thing about open source software like Kodi — there's a huge community of users and coders always working to solve problems like this.)

Find instructions at github.com/CastagnaIT/plugin. video.netflix/wiki/Login-with-Authentication-key and pimylifeup.com/raspberry-pi-netflix.

8. INSTALL THE AMAZON PRIME ADD-ON

Getting the repository for Prime Video requires a different approach. You'll need your PC. First, copy the repository installation file under https://github.com/kodinerds/repo/raw/master/ repository.sandmann79.plugins/repository.

sandmann79.plugins-1.0.3.zip to your PC.

Now connect with the Samba server you previously enabled on your Pi. In your Windows file browser, enter your Pi's IP address, for example: \\192.168.10.18 (those are backslashes); on the Mac we used Finder → Go → Connect to Server → smb://192.168.10.18 (forward slashes). Enter your Samba username and password that you set earlier. Now you'll see the directories that are accessible on your Pi (Figure M). Copy the file *repository.sandmann79.plugins-1.0.3.zip*, which you just downloaded, into the Pi's *Downloads* folder.

On the Kodi main screen, click on the gears symbol to get to the System screen. From the media list, choose Add-ons, then Install from ZIP File. Now, click on the *Home* folder (Figure N), then the *Downloads* folder. You should see the file that you just copied via Samba. Select it

prime
video

Movies and Television Shows for
Prime Members

• [DISCONTINUED] Amazon   Sandmann79 + BlueCop + Romans I XVI - 2.0.8

✓ Amazon VOD   Sandmann79, Varstahl - 0.8.8

Options
Success: You've installed the Amazon Prime add-on! Now just click on it, choose Open, and enter your
username and password for Prime Video.

by clicking on it, then click OK. This makes the repository known to Kodi.

The remaining steps to Install from Repository are the same as for installing the Netflix add-on (above), except that the new repository is called Sandmann79s Repository Leia, and the add-on is called Amazon VOD. When completed, your screen should look like Figure **O**. Click on Amazon VOD and choose Open. On the next screen, click on Connection, then Connect (Figure **P**). Enter the e-mail address and password you used when registering for your Prime subscription.

Go back to Kodi's main screen. You'll now find the Amazon Prime Video symbol in the add-ons section (see title photo on page 56). Click on it, and you can indulge yourself in their selection of things to watch. Enjoy! ⊘

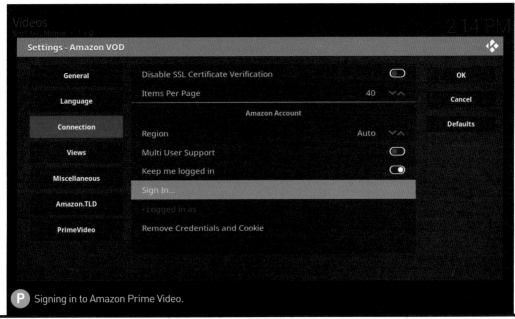

Settings - Amazon VOD

General	Disable SSL Certificate Verification		OK
Language	Items Per Page	40	Cancel
Connection	Amazon Account		Defaults
Views	Region	Auto	
	Multi User Support		
Miscellaneous	Keep me logged in		
Amazon.TLD	Sign In...		
	· Logged in as		
PrimeVideo	Remove Credentials and Cookie		

P Signing in to Amazon Prime Video.

LongHive

Track your beehive's health with this deep-learning, end-to-end solution

Written by Antonio Scala, Evan Diewald, and Nathan Pirhalla

Whether you're an individual hobbyist or a commercial farmer who relies on large-scale pollination, monitoring your hive with simple sensory data can help beekeepers detect problematic trends in colony health. Our project, known as LongHive, is a full-service infrastructure for beehive maintenance, enabled by deep learning (DL) and LoRaWAN communications via the Helium Network. Data-driven beekeepers can install our LongHive system underneath their standard beehives and take advantage of its suite of sensors, pre-trained convolutional neural network (CNN) for classifying the hive's acoustic signatures, and web-based dashboard for easy visualization of the transient signals.

Our goal is to help beekeepers make the most out of their time and reduce the frequency of intrusive hive inspections while still detecting problems within the hive. As opposed to limited-range, power-hungry protocols like Wi-Fi or Bluetooth, the Helium Network's LongFi architecture enables low-power LoRaWAN devices that can operate in a much more remote environment. We combine edge computing and a pre-trained neural network to circumvent the most glaring constraint of LoRaWAN networks — low transaction throughput — in our DL classifier. A Raspberry Pi bears the computational

BEST DEAL

**Subscribe today and get Make: delivered right to your door
for just $39.99! You SAVE 33% OFF the cover price!**

Name _____ (please print)

Address/Apt. _____

City/State/Zip _____

Country _____

Email (required for order confirmation) _____

☑ **One Year**

$39.99   □ **Payment Enclosed**   □ **Bill Me Later**

BOANS3

Make:

Make: currently publishes 4 issues annually. Allow 4-6 weeks for delivery of your first issue. For Canada, add $9 US funds only.
For orders outside the US and Canada, add $15 US funds only.

TIME REQUIRED:
8–10 Hours

DIFFICULTY:
Moderate. Familiarity with Raspberry Pi and deep learning recommended, but not necessary.

COST:
$150

MATERIALS
» **Raspberry Pi 3B single-board computer**
» **Seeed ReSpeaker 2-Mics Pi HAT**
» **Load cells, 50kg (4)** SparkFun Electronics SEN-10245
» **Load cell amplifier** SparkFun HX711
» **Air Quality breakout board** SparkFun CCS811
» **Temperature sensor, DS18B20 type**
» **STM32L0 LoRa/Sigfox Discovery Kit** STMicroelectronics B-L072Z-LRWAN1
» **Wood screws**
» **2×4 board, roughly 8' length**
» **Plywood sheets, ¼"×18"×22"**

TOOLS
» **3D printer**
» **Power drill**
» **Table saw**

ANTONIO SCALA is studying computer science and mathematics at Villanova University. After graduation, he plans to enter the space industry.

NATHAN PIRHALLA has a degree in finance from the University of Pittsburgh. He is also passionate about local farming and making unobtrusive hardware.

EVAN DIEWALD is working on his PhD in mechanical engineering at Carnegie Mellon. He loves to find new applications for deep learning and additive manufacturing.

burden locally, so only the network output (the classification itself) needs to be transmitted over LongFi (Figure Ⓐ).

LongHive Sensor Suite

In our review of relevant literature and existing commercial solutions, we found a slew of passive sensors that have proven to give some indication of hive health. First and foremost, we want to provide beekeepers with real-time data that they will use to augment their existing heuristics and improve productivity.

- Variation in **hive weight** is a sign of honey production and population.
- **Temperature** is a simple but critical source of information; bees like to keep very precise thermal conditions for optimal hive development. In fact, they have fascinating mechanisms for maintaining this delicate homeostasis: when the hive is too hot, they fan their wings to increase convective cooling; when it's too cool, they generate heat by vibrating their flight muscles.
- Similarly, beekeepers must keep an eye on the relative **humidity** in their hive — eggs cannot hatch when it's too dry, but damp conditions can be a sign of mold or disease.
- **Carbon dioxide** is released into the hive as a byproduct of honey production. Thus, a lack of proper ventilation can result in CO_2 poisoning and other maladies. Beekeepers maintain this balance by making tweaks to airflow and insulation.
- The **acoustic signals** emitted by a hive can be a rich source of information, but it will take a more complex processing pipeline to make sense of it (more on this in a moment).

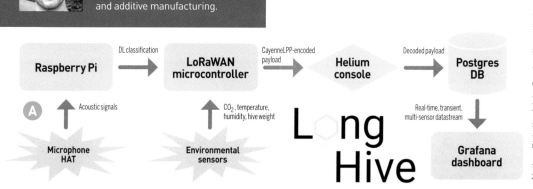

Nathan Pirhalla, Adobe Stock - diyanadimitrova

With this ecosystem in mind, we want to be as unobtrusive as possible. The good news is that beehives have standard dimensions, which means we can design a "one size fits all" solution. If you've ever seen a hive in person, you're probably familiar with the famous stackable assembly. After discussing with a friend who keeps bees, we decided to use the empty hive stand at the bottom to house our electronics, batteries, and load cells (Figure **B**). Wired sensors are threaded up into the hive to give relevant measurements.

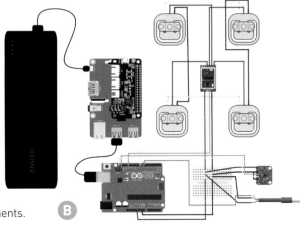

B

Replace this Arduino Uno with the ST LoRa Discovery Kit board, which has the same form factor.

Classifying Buzzing Signals With Deep Learning

While you can tell a lot about a hive's health from first-degree data sources like temperature and humidity probes, researchers have proven that you can also extract useful information by listening to the bees themselves. As a proof of concept, we have implemented a CNN that classifies a hive based on whether or not it has a queen, by encoding the spectral content of its acoustic signals. Once a robust, labeled dataset is collected (hopefully through the LongHive community), we suspect that we can use a similar pipeline to make other classifications.

The training dataset was compiled from an open source publication (kaggle.com/chrisfilo/to-bee-or-no-to-bee), where beekeepers recorded their hives and labeled the audio files according to whether or not they had a queen. Because it represents a variety of geographic locations, recording techniques, and background noise, the data is robust and generalizable. We split the WAV files into 4.5-second segments, resulting in about 2,000 training samples per class (queen or no queen).

In a purely temporal domain, these acoustic signals are not easily separable, as it is difficult (for a DL model) to differentiate audio of differing amplitudes and background noise. *Mel spectrograms* are commonly used for audio classification, as they extract relevant spectral information from the time-series signals into an image (Figure **C**), allowing us to take advantage of mature CNN-based techniques. The X-axis is time, the Y-axis is frequency, and the color is the

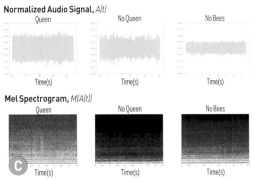

Normalized Audio Signal, *A(t)*

| Queen | No Queen | No Bees |

Time(s)   Time(s)   Time(s)

Mel Spectrogram, *M(A(t))*

| Queen | No Queen | No Bees |

C Time(s)   Time(s)   Time(s)

To increase the separability of the dataset, we transform the raw audio signals (top row) into the time-frequency domain. Here we see the spectrogram outputs for (bottom row, left to right) hives with a queen, hives with no queen, and a control case.

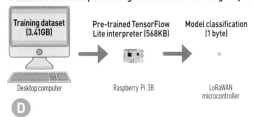

Data Distillation: representing vast database in a single byte

Training dataset (3.41GB) → Pre-trained TensorFlow Lite interpreter (568KB) → Model classification (1 byte)

Desktop computer   Raspberry Pi 3B   LoRaWAN microcontroller

D

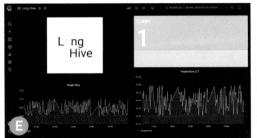

E

Nathan Pirhalla

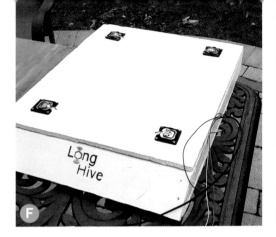

power of the signal at that frequency band.

Once the mel spectrograms were cropped and resized to 256×256×3 inputs, they were fed into the CNN. We found network training to be somewhat unstable, likely due to the small and noisy dataset. (For you machine learning fans, the architecture contains about 144,000 trainable parameters — for reference, the groundbreaking AlexNet architecture has over 60 million parameters! — and consists of descending convolutional and max pooling layers with Leaky ReLU activations for feature extraction and two fully connected layers for classification.) We wanted to keep the size as small as possible, in order to run optimally on the Raspberry Pi. The final accuracy for this binary classification (queen/no queen) was 89% on a test set, but as the LongHive community grows, the model will only improve.

Edge Computing and Grafana Dashboard

For real-time model evaluation, it's computationally inefficient to run a full-blown TensorFlow implementation on the Raspberry Pi, so we're using ARM-friendly TensorFlow Lite for the classification task. The pre-trained TF model was exported, its architecture and weights converted into *.tflite* format, and copied to the Pi's local memory. To collect the audio signals, we're using the ReSpeaker 2-Mics Pi HAT, which has a well-documented Python library. We're also using the same exact pre-processing pipeline to generate the mel spectrogram test images as we used for the training data. Saving the recording, calculating the Fourier transforms, and evaluating the model takes about 10 seconds on the Pi. The classification label (**1**: queen

detected, **0**: no queen detected) is transmitted to the STM microcontroller board via the serial port at predefined intervals.

This is edge computing at its finest: we distilled several gigabytes of training data down into a pre-trained 500KB TFLite model that can be loaded into the Pi's RAM. Upon evaluation, all this knowledge is characterized in the classification by *a single byte* (Figure **D**). LongFi may be known for its low throughput, but that doesn't mean it can't represent vast amounts of information.

For real-time analysis, the transient sensor data can be viewed from any web browser on our Grafana dashboard (Figure **E**). Grafana provides a modular, professional-looking medium for displaying a wide range of data types.

Future Plans

Our project was fortunate enough to win the Grand Prize of the #IoTForGood challenge on Hackster.io. We plan to invest much of our prize winnings into continued development and refinement of our prototype (Figures **F** and **G**). This iteration is a viable proof of concept of the potential of the LongHive system, but it's far from optimal. The Raspberry Pi and LRWAN-1 board will eventually be replaced by specialized hardware with deep sleep capabilities for improved battery life performance. In our prototype, we used a USB battery pack to power the system for three days, but we believe we can increase longevity by several weeks by refining the electronics and code. ⬤

Get the code, build files, and even more details at hackster.io/354300/longhive-12d952

Tin Can Rocket Stove

Get cooking with this ultra-efficient woodstove made from scrap

Written by Daniel Connell

TIME REQUIRED:
30–60 Minutes

DIFFICULTY:
Easy–Intermediate

COST:
$0–$5

MATERIALS
» **Empty steel food cans: large (3), medium (2), and small (3)** approximately 2,200g, 800g and 400g, respectively
» **Twisting wire, about 60cm** or fencing wire
» **Perlite, vermiculite, pumice, or wood ash, about 5 liters** for insulation

TOOLS
» **Can opener** The lever-hook type, like on a pocketknife or multitool, works well.
» **Tinsnips**
» **Blunt-nose pliers**
» **Marker pen**
» **Short stick**
» **Work gloves** The process involves moderately sharp metal.
» **Gaffer tape or duct tape (optional)**
» **Protractor (optional)** or other angle gauge
» **Drill (optional)** for starting cuts

Here's how to make a highly efficient, cheap, and easy-to-build rocket stove, optimized for cooking in off-grid environments. It can be made in less than an hour from a couple dollars' worth of easily sourced, mostly scrap materials.

This is a good option for homesteaders, preppers, survivalists, and anyone camping or living off grid, but it's also useful and fast to deploy in disaster areas and refugee camps.

BUILD YOUR ROCKET STOVE

Before you begin, please watch my assembly video at youtube.com/watch?v=ywljr9RKExQ, and put on your work gloves.

1. SHAPE THE COMBUSTION CHAMBER

Take one of your medium sized cans (Figure ) and use a can opener to cut through the bottom, about three quarters of the way around the inside circumference.

Draw a curved line that starts and ends at each end of that cut, coming in at roughly a 30° angle, up the wall of the can. Then draw two vertical lines that divide your curved lines into rough thirds (Figure ).

Pull the can bottom outward a little so it's out of the way, and cut the two lines up to the level of the curve.

Push that center tab of metal you just cut into the can, and push the can bottom down onto that.

Use pliers to bend the two outer, triangular tabs down to lock everything in place (Figure ).

2. CUT THE LOWER INSULATION JACKET

Take a large can, and place the 30° angled base of your medium can against it, centered about halfway up the wall. Draw a curve around the medium can onto the large one (Figure ).

Draw a star of lines across the resulting oval to divide it into 12 sections.

Use the can opener or tinsnips to punch a hole in the center of the star, then use tinsnips to cut all the lines, being careful not to cut your hands on the metal. Bend all 12 sections out with pliers (Figure ).

Check that the medium can will fit within this hole, sloping down/out at a 30° angle. Enlarge the hole if needed, but keep it tight; you don't want more of a gap than necessary.

Tristan Copley Smith

Rocket Stove: How It Works

Rocket stoves burn hot but consume only half the fuel of other woodstoves, for two key reasons: efficient combustion, and efficient direction of heat. This makes them healthier too: They emit little or no smoke, and far less particulate pollution and carbon monoxide.

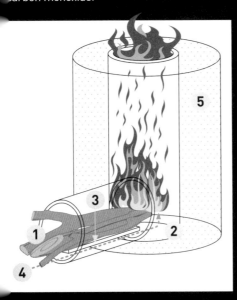

1. **Small-diameter fuel** — sticks burn more efficiently than large pieces of wood, and extinguish quickly when cooking's done
2. **Combustion chamber** — elbow-shaped *L-tube* for horizontal fuel feed and vertical heat direction; aka *burn tube*
3. **Fuel shelf** — admits small fuels only, allows airflow beneath
4. **Air gap** — provides high air-to-fuel ratio for efficient combustion
5. **Insulated chimney** — increases air draft and directs heat to cookware; confines wood gases and smoke in high heat for near-total combustion.
   —*Keith Hammond*

3. ATTACH THE CHIMNEY

Take a small can and remove the top and bottom with the can opener. Place the small can on top of the combustion chamber (medium can) at a 30° angle (on the intact side, away from the bent tabs), and draw around it (Figure **F**). Again mark, cut, and bend out 12 slice sections to open up the hole.

Insert the small can into this hole so its base is just within the wall of the medium can, not protruding into it more than necessary. Place a loop of twisting wire around the outside of the triangle slices. Use pliers to twist the wire tight so that it holds the small can in place (Figure **G**).

Bend down the tips of the triangles so they lock the wire in place.

4. MAKE THE FUEL/AIR FEED TUBE

Take the remaining medium can and, with the can opener, make a cut about two thirds of the way around the inside circumference of the can's base, and then another smaller cut, leaving 2cm gaps of uncut metal in between.

Bend the resulting larger flap of metal outward to about 90° or so, to make a little shelf for your pieces of wood to rest on.

The bottom cut is an air gap, to let in air to feed the base of the fire. You can bend it inward slightly to open it.

Use the can opener to cut out and remove the top of the can if it's not already. With the tinsnips, make a 1cm cut in the wall of the can opposite the shelf, and bend the wall inward a little (Figure **H**) so that the can will fit inside the other medium can (the combustion chamber).

5. ASSEMBLE THE STOVE

Place the combustion chamber into the large can so that the opening of the medium can protrudes slightly through the star cut in the large can.

Then take the shelved feed tube and force it into the front of the combustion chamber so that their outside and inside lips of metal lock together.

Twist the feed tube so that the shelf is horizontal, with the air gap slit at the bottom (Figure **I**).

6. FINISH THE CHIMNEY

Remove the tops and bottoms from the two remaining small cans. Cut 1cm slices into one end of each can.

Connect these cans to each other in the same way as previous.

Then connect these two to the open top of the small can in the combustion chamber assembly (Figure 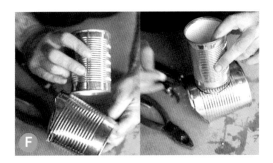 on the following page).

5. FINISH THE OUTER JACKET

Remove the tops and bottoms from the remaining two large cans.

Make four 1cm cuts evenly spaced around one end of each. Bend two of the resulting tabs inward slightly, and the other two outward slightly.

Drop first one and then the other onto the top of the assembled large can, so that the inward tabs are inside of the can beneath, and the outward tabs are outside.

The three large cans are now stacked and connected to each other. The stack is strong in terms of supporting downward force from heavy pots set on top of the stove, but it's not strongly locked together, so be careful when you pick the stove up that it doesn't come apart. The insulation should keep the outside surface from becoming too hot, so you may want to use tape or rivets to attach the large cans, but if the connections are tight enough then this shouldn't be necessary.

Tristan Copley Smith, Adobe Stock - artinspiring and Oqvector

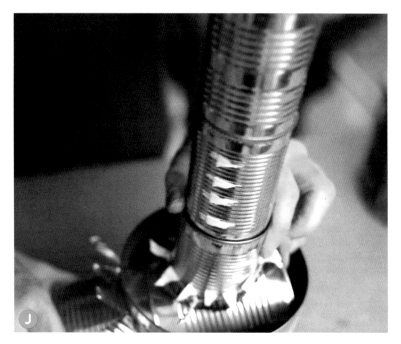

DANIEL CONNELL is an open source low-tech alternative infrastructure designer. He designs and tutorializes things that people can make from recycled materials to generate their own energy, purify water, cook, and communicate. facebook.com/ opensourcelowtech. org

Twist a loop of wire around the star sections holding the protruding feed/burn tube so it's held in place at about a 30° downward angle (Figure Ⓚ). This is so that the coals drain forward, don't clog the back of the burn tube, and burn the full length of the wood, producing more flame and increasing the efficiency.

Fold the tips of the triangles back so they lock off the wire and are less of a hazard.

6. INSULATE YOUR ROCKET STOVE

Start filling the space between the small and large cans with perlite, or whatever you're using as insulation. Keep everything aligned, with the medium cans on their 30° downward angle and the small cans centered in the large, while you use a short stick to push down and compress the insulation as you go.

> **TIP:** Use one of the large can's removed lids as a shield over the open top of the small cans, to direct the insulation into the gap and not down the chimney.

7. CUT THE POT SUPPORTS/FLAME GAPS

Finally, make four cuts of about 3cm–5cm into the top edge of the outer jacket, and fold down triangles of metal on each side of all of them (Figure Ⓛ). This provides exit gaps for the flame,

Ask questions and post your build at facebook.com/groups/opensourcelowtech, and find more low-tech projects at opensourcelowtech.org.

Tristan Copley Smith

and reinforces the resulting tabs so that they can hold larger weights without folding.

FEEL THE BURN

It's important the first time you light your rocket stove to do so outside, in a ventilated area, as that first fire will burn off all the plastics and other chemicals in the cans. After about 10–15 minutes of running at full temperature, that stuff should all be gone and the stove can be used in more enclosed spaces thereafter.

> **CAUTION:** Obviously do not use the stove anywhere that the fire could spread or be hazardous if it were to be knocked over or otherwise spill flame and burning material.

FEEDING FUEL

The best fuel for your rocket stove is dry, untreated wood, a couple of centimeters square and about 20cm–30cm long. This can be split kindling, or sticks and small branches.

Ideally you only want to have two or three pieces of wood in the stove at a time, as overfilling it will reduce the space for airflow. The ideal configuration is where the tube is about one-quarter full of hot coals (once it's been running for some minutes to produce them) and one-quarter

full of wood that's fully ablaze, leaving half the space for air, as this will produce maximum flame, temperature, and efficiency (Figure Ⓜ).

Keep feeding the wood toward the back as it burns, adding a new piece when it's fully burned away.

LIGHT IT UP

To light, place some paper or other tinder in the burn tube, followed by a few small sticks or splinters of kindling, followed by two or three larger pieces. It's usually easiest to then light the tinder with a lighter or match through the air gap at the bottom.

The stove will produce smoke for the first couple of minutes, but this will reduce to almost zero as it comes up to peak temperature. If you see smoke after that, it usually means the fire is burning out and needs more fuel.

The stove should use about 80% less wood than an equivalent open fire, and produce greatly less smoke and emissions. The coals will burn completely to ash, and what little is left behind can be dispersed by blowing into the burn tube. I haven't found cleaning to be much of an issue. ◐

The Open Book
and E-Book
the
FeatherWing

Solder your own e-reader for books and texts in every language

Written and photographed by Joey Castillo

TIME REQUIRED:
1–2 Hours

DIFFICULTY:
Intermediate

COST:
$90–$120

MATERIALS
See the complete interactive BOM at kitspace.org/boards/github.com/joeycastillo/the-open-book/ebook-wing

- » **Custom circuit board, E-Book Wing PCB** tindie.com/products/19994
- » **Adafruit Feather M4 Express microcontroller board** adafruit.com/product/3857
- » **E-paper display, 4.2", 400×300 pixels** Good Display #GDEW042T2
- » **Resistors: 0.47Ω (1), 100Ω (2), 1kΩ (2), and 10kΩ (8)**
- » **Capacitors, ceramic: 4.7µF (1), 1µF (12), 10µF (2), and 100µF (1)**
- » **Zener diodes, 3.6V 500mW (2)**
- » **Schottky diodes, 30V 500mA (3)**
- » **Ferrite beads, 1kΩ (2)**
- » **Power inductor, 10µH 1.3A 170MΩ**
- » **Transistor, MOSFET N-channel 100V 1.6A**
- » **I²C interface IC chip** for GPIO expansion
- » **Memory IC chip, NOR flash**
- » **Memory IC chip, serial SRAM, 256K**
- » **Headphone jack, 3.5mm**
- » **MicroSD card slot**
- » **Female headers** for connecting the Feather
- » **JST-PH connectors: 4-pin (1) and 3-pin (2)** aka STEMMA ports
- » **Various connectors, buttons, and switches** see BOM for details
- » **Thin solder**
- » **Flux**
- » **Solder wick**

TOOLS
- » **Soldering iron, fine tipped**
- » **Tweezers, fine tipped**
- » **Multimeter**
- » **Magnifying glass, loupe, or microscope**

JOEY CASTILLO is a Brooklyn-based maker and software engineer.

Handheld gadgets surround us, but often these slabs of glass and silicon feel like they arrive fully formed in our lives. We don't often get the chance to understand how, say, a Kindle works — or even better, how we might make one ourselves! The Open Book Project aims to create a simple, comprehensible device for reading texts in all the languages of the world — an e-reader that can be built by hand with the tools in the average maker's workshop.

The E-Book FeatherWing is the simplest version of the build. Functionally it's a "wing" — an accessory board that pairs with an Adafruit Feather M4 Express, which provides the microcontroller and the battery charging circuitry. Almost all of the E-Book wing's surface-mount parts are reasonably large and easy to solder by hand, which makes it a great project for getting started with surface-mount soldering.

In this guide, you'll build your own E-Book FeatherWing. With its 4.2" e-paper screen and seven clicky buttons on the front, it's a great device for reading, but it's also a highly capable platform for doing more! You can use it to play sound from the headphone jack, display sensor data from the trio of STEMMA ports, or even stack an AirLift FeatherWing to connect to the internet.

1. PREPARE FOR THE BUILD
You'll need the E-Book FeatherWing printed circuit board. You can order one on Tindie, or send the board design files to a PCB fabrication house like OSH Park.

You'll also need to order the parts that go on the board. Almost all of these parts are available at Digi-Key; you can use the one-click bill of materials, or BOM, at Kitspace to order almost all of them in one fell swoop. The screen will come direct from the manufacturer, Good Display.

Make sure you have the Arduino IDE installed on your computer, along with Adafruit's SAM Board support package, which will support the Feather M4 Express. In addition, make sure you have installed the following Arduino libraries:
- Adafruit GFX
- Adafruit BusIO
- Adafruit MCP23008
- Adafruit EPD
- Adafruit SPIFlash

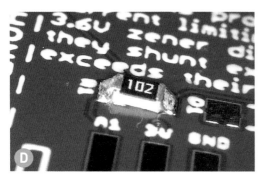

- SDFat — Adafruit Fork
- arduino-menusystem
- The Open Book
- Babel

When you have your development environment set up and all your parts on hand, set up your work area. Print a copy of the BOM so you can cross out parts as you solder them to the board. It also helps to have a printout of the backside of the board, in case you can't read a part number on the silkscreen. Make sure to have your solder wick and flux handy, in case you need to do any rework. Finally, before you build, watch my E-Book FeatherWing assembly video at github. com/joeycastillo/The-Open-Book.

2. SOLDER THE PASSIVE COMPONENTS

If you've never soldered tiny surface-mount components before, don't worry! It's more straightforward than you might think. Let's start with the resistors and capacitors: all of these parts are 0805 sized, which is small but still big enough to solder by hand with just your eyes or some reading glasses.

Find the first part you want to solder on the board. Let's say it's R1. First, put down a blob of solder on one of the two pads (Figure Ⓐ).

Next, using your tweezers, position the part near that blob of solder. Heat up the solder, and then move the part into that spot. This will tack it into place (Figure Ⓑ).

Finally, apply solder to the other side of the part (Figure Ⓒ).

That's it! R1 is placed (Figure Ⓓ).

Repeat these steps for each of the resistors and capacitors in the BOM, as well as inductor L1.

You can use a very similar strategy for the diodes D1–D5, but note that the diodes need to be mounted in a specific direction. Each diode has a small gray line on its plastic body that should face the same direction as the line in the diode symbol on the silkscreen (Figure Ⓔ).

There are also two different kinds of diodes; the two Zener diodes need to go in the Extra Ports block, while the three Schottky diodes need to go in the E-Paper Display block.

While we're here, we can also solder the

surface mount buttons! It's identical to the way you soldered the resistors and capacitors: put a blob of solder on one side, set the button in place, then secure the other side (Figure F).

You can use this same technique to solder MOSFET Q1 into place. That part has three pins, so just make sure each pin matches with a pad.

3. SOLDER THE INTEGRATED CIRCUITS

There are three ICs on this board: a flash chip, an SRAM chip and a GPIO expander. Like the diodes, when soldering these parts you'll need to be aware of the orientation. Each part has a small dot on its plastic body that should match with the dot on the board's silkscreen (Figure G).

The strategy is similar to what we did for the diodes: first, put a blob of solder on one of the pins at the corner (Figure H).

Then, place the IC while heating up that bit of solder (Figure I).

Before you solder any other pins, look at the alignment of the chip: all the pins should be aligned with all the pads. If they're not, reheat the pin you soldered, and move the chip around until it is aligned. Then solder the pin at the opposite corner to lock the IC into place.

If the placement looks good, you can solder the rest of the pins (Figure J).

If you notice solder connecting some neighboring pins, don't worry! This is called *bridging*, and it's easy to fix. If it's not a lot of solder, you might be able to fix it by simply adding some flux and reheating the pins (Figures K, L, M, and N). The flux helps the solder flow to the metal pins and nowhere else.

If that doesn't work, you can place your solder wick on the bridged pins and heat it to soak up the excess solder (Figures O and P).

4. SOLDER THE SD CARD SLOT

This part might look tricky, but it's not so bad! First, like last time, we're going to apply a blob of solder to one of the mounting pads, and heat it up while we maneuver the microSD slot into place (Figure Q).

This microSD socket has a nice big window on top, which should make it easy to see if all the pins are aligned (Figure R on the following page).

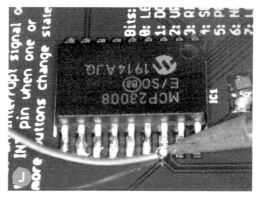

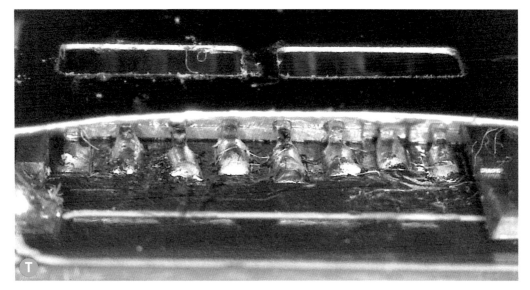

If not, heat up that mounting pad, and move the slot around until the socket is aligned. Then solder the remaining mounting pads.

The trick with this card slot is to feed your thin solder in from the top (Figure S), and insert your soldering iron from the side. It helps to use a PCB holder, or to simply place the board at the edge of a table while you work, so that you can insert the tip of your soldering iron straight in.

Use your magnifying loupe to look inside the slot and make sure all the pads are soldered in place (Figure T).

5. SOLDER SURFACE-MOUNT CONNECTORS

There are seven surface-mount connectors to solder on this board: the headphone jack, three STEMMA ports, two Feather headers, and the 24-pin flex connector.

The headphone jack is easiest: it has two plastic board guides that match the two holes on the PCB. When you set it on its footprint, it will fall into place. Just apply solder to all four pads.

The STEMMA ports are similar to the ICs: you can place some solder on one of the mounting points, move the port around until it matches the outline on the silkscreen, and then solder the remaining pads. Make sure the 4-pin port goes in the middle, with the 3-pin ports on either side!

Same with the Feather headers: tack one corner into place, then the other corner, and check the placement of both headers before you solder the rest of the pins. (It helps to gently rest your Feather M4 on top to make sure it's going to fit.)

Finally, the 24-pin connector. This is the trickiest part on the board, no doubt. But with your solder wick and flux, you'll have it placed in no time! First, place a blob of solder on one of the two large mounting pads on either end of the connector. Then heat it up, and maneuver the connector into place (Figure U).

Using your magnification loupe, make sure all

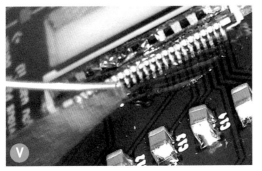

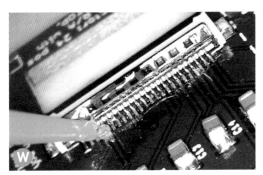

24 pins are aligned. It's pretty easy to see when they are: you should see the blue silkscreen visible between each set of pins. Then, drag your soldering iron and solder across the pins, and solder those pins into place! It can help to put some extra flux down, just to make sure the solder goes where you want it to go (Figure V).

You're almost inevitably going to have some bridged pins here. That's okay! Clean your soldering iron, apply some flux (Figure W), and drag the bridged spot away from the connector (Figure X). If you have a larger bridge, it can help to use your solder wick here too.

It may take a few tries, but once you see that there are no solder bridges and all the pins are soldered to pads, you're almost done!

6. TEST THE BOARD

Just to be extra safe, use your multimeter to check for shorts between the 3V and GND nets. Put your multimeter in continuity mode, and touch the two probes to the 3V and GND pads in the Feather header. If it beeps, inspect the board for any solder bridges that may be connecting power and ground nets. If it doesn't beep, you're good to plug in your Feather M4!

Carefully take the fragile e-paper display out of any protective packaging it's in. Thread the screen's flex cable through the hole up top, and secure it in the flex connector (Figure Y).

Taking care to avoid damaging the screen, turn the device over so you can see the screen, and power up the Feather.

Load the *Open_Book_Screen_Test* sketch from the Open Book examples. Make sure that your Feather M4 is selected, and then run the sketch! You should see the screen flash to life, and display the Open Book Project logo.

7. SOLDER THE FINAL PARTS

Unplug the Feather from power, disconnect the screen's flex cable from the 24-pin connector and set the screen aside.

There's one solder jumper we need to close to make the flash chip work. Locate the BCS solder jumper, near the top of the long Feather header, and connect both sides with a blob of solder.

Next up: buttons! Insert each of the through-hole buttons so they face the front of the device, and solder them in place from the back.

Then insert the through-hole slide switch from the back of the device, and solder it in place from the front.

Finally, thread the screen's flex cable through the hole one last time. Insert it into the 24-pin connector, and secure it in place. Using double-sided tape, secure the screen to the front of the PCB. Congratulations, your board is assembled (Figure 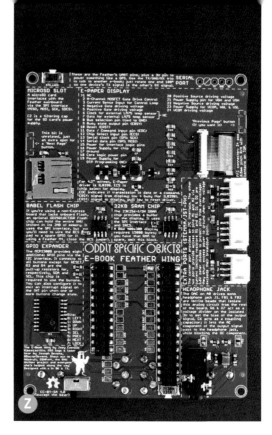)!

8. BURN THE BABEL FLASH CHIP

The secondary flash chip on the E-Book FeatherWing is dedicated to font and language support, but it currently doesn't have any data on it. When you downloaded the Babel library, you also downloaded a two-megabyte BLOB file containing information about how to display all the world's languages. We're going to copy that blob to the microSD card, and then run a sketch to burn that image to the flash chip.

Locate the *babel.bin* file; it's in the *babelconvert* folder inside the Babel library you downloaded (Figure 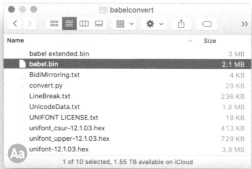). Copy *babel.bin* to your microSD card, and eject the drive. Then plug the microSD card into the socket on the E-Book FeatherWing. Run the *BurnBabelBurn* sketch from the Babel examples, open the Serial Monitor, and follow the instructions to burn the font data to the chip (Figure **Bb**).

If the image verifies, you're done! You can delete the *babel.bin* file from the SD card; the Babel chip will retain this data indefinitely.

9. READ A BOOK!

Finally! The book reading software, so far, is pretty basic: We're using MVBook to read text files that have been converted to the MVBook format (although we hope to have plain *.txt* files be the format in the future).

In the MVBook example folder, you should find a file called *books.zip*, which is a small selection of public domain works. Copy the contents of this zip file to an SD card, and insert it in the E-Book FeatherWing.

Go to File → Examples → Open Book and run the *Open_Book_MVBook* sketch! You should see a list of titles and authors, and a selection indicator on the left (Figure). Use the up and down arrows to select an item, and the center Select button to pick one. In book reading mode (Figure **Dd**), use the far left and right Previous Page and Next Page buttons to flip through pages. Use the Select button to return to the main menu.

EXPAND YOUR E-BOOK FEATHERWING

The E-Book FeatherWing is useful for reading books, but it can do quite a lot more!

- The screen supports 2-bit grayscale mode, which is great for displaying photographs, and the Feather M4's SAMD51 is powerful enough to decode JPEG images on the fly!
- The E-Book FeatherWing can stack with several other FeatherWings, meaning you can add a Wi-Fi connection (AirLift wing), real-time clock (DS3231 wing), or even GPS capabilities (Adafruit Ultimate GPS wing)!
- The STEMMA ports on the side open up many possibilities for everything from sensors to speakers to NeoPixels.
- You can also add a 400mAh battery, Adafruit #3898, to take your book on the go.

LEVEL UP TO OPEN BOOK FEATHER

If you're up for a more advanced build, The Open Book Feather board brings the SAMD51 onto the main board, and adds features like stereo audio output, microphone input, and more free pins for supporting other wings.

Both boards support CircuitPython, Adafruit's awesome branch of MicroPython that's great for education and experimentation. ◑

Cc

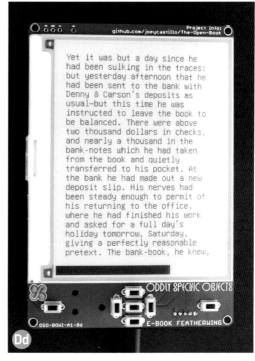

Dd

For more information about the Open Book, including how to make a 3D-printed or laser-cut enclosure, check out the Open Book's GitHub repository at github.com/joeycastillo/The-Open-Book.

E-Felted Fantastical Mushroom

Practice "slow making" and mindfulness while needle-felting, then sew on a light to complete the magical experience

Written by Agnieszka Golda and Jo Law

MATERIALS
With some eco-maker tips on getting them sustainably:

» **Wool roving or sliver, 12" (30cm) lengths, in 4 colors** We're using 100% pure New Zealand Corriedale wool sliver; merino wool top is too fine for this project. Try to source local, sustainably produced, non-mulesed wool. You can dye your own wool and felted objects using eco-friendly natural dye processes.

» **LED sequin**

» **Uncoated wire, bare silver or copper, 50cm length, 32–30 gauge (0.2mm–0.25mm)** You can use copper wire stripped from an old audio cable.

» **3V battery** We're using a 3.7V lithium polymer (LiPo) battery. You can use coin, AA, or AAA batteries, but rechargeable is ideal.

» **Matching battery holder with a switch**

» **Alligator clips (optional)** to connect wires to battery

TOOLS
» **Felting foam or mat Use** 100% wool felted mats or plant-based (corn or soy) foams.

» **Felting needle, 38 gauge, star or triangle shape** Both needles work well for bulk sculpting work and finer detail.

» **Sewing needle**

» **Scissors**

» **Single needle holder (optional)** Try a holder made from biodegradable and sustainably harvested materials, such as wood.

» **Finger cots/protectors (optional)** You can make your own finger cots by upcycling old rubber or leather gloves.

AGNIESZKA GOLDA PH.D.
and **JO LAW PH.D.** combine their expertise in textiles, design, visual and media arts, and bring together art, science, and technology to develop creative strategies for improving people's and the planet's well-being. Their eco-making workshops for the Global Challenges: Future Makers Project at University of Wollongong explore how creativity and science can contribute to building environmentally sustainable and healthy living futures.

The outbreak of Covid-19 is having an enormous and diverse impact on people globally. Many people feel isolated, frustrated, and overwhelmed. To manage our well-being, crisis support organizations emphasize a "calm approach" that incorporates creative, mindful practices in our daily routines. Creative activities can help us to express our feelings and to connect with family, friends, and communities. Today, the connection between well-being and engagement with "slow" approaches to making is more meaningful than ever.

In this project, we show you how to create an E-Felted Fantastical Mushroom — an e-textile sculpture that involves needle-felting and electronics. The project deliberately explores a slow and contemplative approach to creating that takes into account the life-cycle and sustainability of materials. By felting soft wool to create a fantastical form that lights up, we can calm the mind, encourage mindfulness, and express feelings of affection and connectivity with our communities and the natural environment.

1. Separate the fibers
To divide a length of roving or sliver, hold your hands apart about 6" (15cm) and gently pull. The fibers will separate easily without breaking (Figure Ⓐ). If your hands are too close together the fibers will not separate, and they are likely to break.

2. Felt the mushroom cap
Place your sliver on the felting mat. Roll the sliver

Ⓐ

Jo Law

tightly (Figure **B**). As you roll, tuck in both ends (Figure **C**).

Before you reach the end, about 2" (5cm) away, poke across the surface with your needle several times to secure the fibers (Figure **D**). Felt gently through the fiber, just touching the surface of the mat, to prevent the cap from felting into the mat.

> **TIP:** Extend the life of your needles by preventing needle breakage. Poke gently into the mat: just touch the surface, and don't punch deeply into it.

Roll and fold the end of the sliver. Felt the surface a few times to secure the fibers (Figure **E**). Fold in half to form a round form (Figure **F**).

Poke the surface from all directions by rotating the form (on the top, horizontally around the edges, and underneath). The cap will become denser and smaller. Continue the process to form a cap shape (Figure **G**). You can vary the size of the cap by wrapping more wool and felting it around the cap .

To measure for appropriate compression and firmness, gently pinch the surface to make sure that the fibers do not come away easily. Then, squeeze the form (Figure **H**) and compare its firmness using the "fingers/thumb test."

FINGERS/THUMB TEST:
Hold your thumb and index finger together, and feel the firmness of your thumb muscle. It's soft; if the firmness of the cap feels similar, then more felting is needed. Now hold your thumb and ring finger and then your little finger together, and feel your thumb muscle again. It's firmer; if the firmness of the cap feels similar, then felting is complete.

3. Felt the mushroom stem
Arrange some sliver into a four-finger width (Figure **I**). Roll the sliver tightly into a log (Figure **J**). As you roll, tuck in the ends.

Poke across the surface of the stem with your

B

C

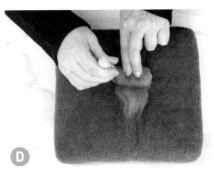

D

E

F

Jo Law

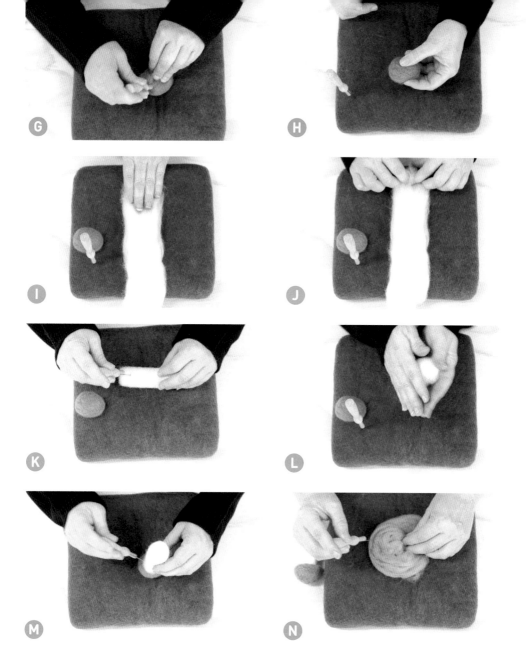

needle to secure the fibers (Figure Ⓚ).

You can roll the stem gently between your fingers to tangle the fibers further (Figure Ⓛ). Continue to felt around the stem. Leave one end a little bit softer. When ready, apply the fingers/thumb test to the rest of the stem.

Attach the softer end of the stem to the mushroom cap. Poke underneath the cap, around the stem, and from the top to secure the cap (Figure Ⓜ). Tug gently to make sure the cap is firmly attached.

4. Felt the moss ground

Arrange the fibers into an organic shape (Figure Ⓝ). Poke on top with your needle to secure the fibers. Turn the shape and work on the back.

Add a second color. Felt the layers together and make the base firm (Figure Ⓞ on the following page). Firm up the edges to a desired shape.

Attach the mushroom stem to the moss (Figure Ⓟ, next page). Work around the sides and from underneath to secure the mushroom. Tug gently to ensure that the mushroom is firmly attached.

O

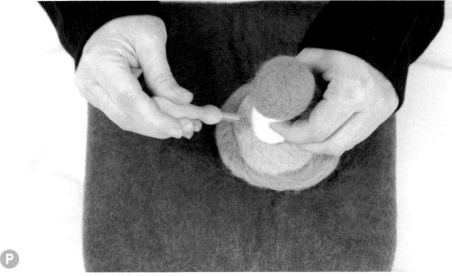

P

Q

R

5. Sew on the electronics

Start by wiring up your LED sequin. These LEDs are designed to be sewn onto soft fabric (Figure Q). They come with two holes for sewing. One is the positive (+) terminal and the other is the negative (–) terminal.

For this project, we will sew with uncoated wire (Figure R) — don't use beading wire or magnet wire for this. Cut two lengths of wire, each about 4" (10–12cm) long. Thread a wire about ¾" (2cm) through one hole, bring the end back around and twist it around the standing part of the wire, and wrap tightly. Repeat with the wire in the other hole (Figure S).

At this point, you can check your connections. The LED sequin has a positive (+) mark and a

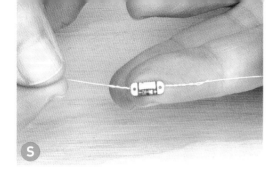

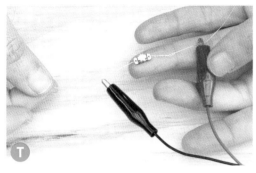

negative (–) mark next to the holes. Connect your wires to the battery using your alligator clips: positive to positive, negative to negative (Figure T). The LED will light up. If not, check your connections.

Now decide where you'd like to put the LED on your felted mushroom. Thread your wire through a sewing needle and sew the wire onto the mushroom (Figure U). Make sure the wires don't touch each other. Sew the negative and positive wires at opposite sides of the mushroom stem. Sew the wires through to the base, with each end poking out at opposite sides (Figure V).

Connect the wires to your battery (Figure W). You can create a more secure connection by sewing your wires directly to a battery holder and attaching the holder to the base of the mushroom.

A Shroom in Your Room

Place your e-felted mushroom in your workspace or your living room, or take it with you to a special spot in your garden at night to create a magical space for contemplation and mindfulness. Make one for a friend to express your affection and connection at this time of physical distancing.

Add sensors, switches, or an Arduino to the circuit and turn your e-felted mushroom into a sensor-triggered nightlight or a programmable porch light. Add a small 3V solar panel to recharge your battery to make this project even more eco-friendly (Figure X). ⊘

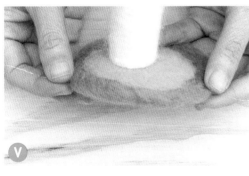

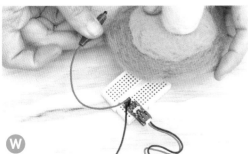

Machine Learning on Microcontrollers

Put real live artificial intelligence into your projects with TinyML and Arduino

Written by Helen Leigh

TIME REQUIRED:
2–3 Hours

DIFFICULTY:
Easy

COST:
$40–$50

MATERIALS
» **Arduino Nano 33 BLE Sense microcontroller board** with USB cable

TOOLS
» **Computer with Arduino IDE software** free at arduino.cc/downloads

HELEN LEIGH is a hardware hacker who specializes in music technologies and craft-based electronics. Say hello to her on Twitter @helenleigh.

Arduino, Adafruit

Machine learning is getting lots of attention in the maker community, expanding outward from the realms of academia and industry and making its way into DIY projects. With traditional programming you explicitly tell a computer what it needs to do using code; with machine learning the computer finds its own solution to a problem, based on examples you've shown it. You can use machine learning to work with complex datasets that would be very difficult to hard-code, and the computer can find connections you might miss!

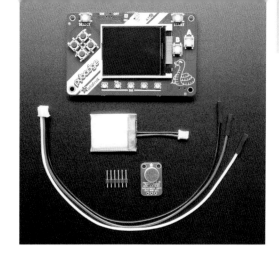

HOW MACHINE LEARNING WORKS

How does machine learning (ML) actually work? Let's use the classic example: training a machine to recognize the difference between pictures of cats and dogs. Imagine that a small child, with an adult, is looking at a book full of pictures of cats and dogs. Every time the child sees a cat or a dog, the adult points at it and tells them what it is, and every time the child calls a cat a dog, or vice versa, the adult corrects the mistake. Eventually the child learns to recognize the differences between the two animals.

In the same way, machine learning starts off by giving a computer lots of data. The computer looks very carefully at that data and tries to model the patterns. This is called *training*. There are three ways to approach this process. The picture book example, where the adult is correcting the child's mistakes, is similar to *supervised machine learning*. If the child is asked to sort the pages of their book into two piles based on differences that they notice themselves — with no help from an adult — that would be more like *unsupervised machine learning*. The third style is *reinforcement training* — the child receives no help from the adult but instead gets a delicious cookie every time they get the answer correct.

Once this training process is complete the child should have a good idea of what a cat is and what a dog is. They will be able to use this knowledge to identify all sorts of cats and dogs — even breeds, colors, shapes, and sizes that were not shown in the picture book. In the same way, once a computer has figured out patterns in one set of data, it can use those patterns to create a *model* and attempt to recognize and categorize new data. This is called *inference*.

MOVING TO MICROCONTROLLERS

ML training requires more computing power than microcontrollers can give. However, increasingly powerful hardware and slimmed-down software lets us now run inference using models on MCUs *after* they're trained. *TinyML* is the new field of running ML on microcontrollers, using software technologies such as TensorFlow Lite.

There are huge advantages to running ML models on small, battery-powered boards that don't need to be connected to the internet. Not only does this make the technology much more accessible, but devices no longer have to send your private data (such as video or sound) to the cloud to be analyzed. This is a big plus for privacy and being able to control your own information.

BIAS BEWARE

Of course, there are things to be cautious about. People like to think computer programs are objective, but machine learning still relies on the data we give it, meaning that our technology is learning from our own, very human, *biases*.

I spoke to Dominic Pajak from Arduino about the ethics of ML. "When we educate people about these new tools we need to be clear about both their potential and their shortcomings," Pajak said. "Machine learning relies on training data to determine its behavior, and so bias in the data will skew that behavior. It is still very much our responsibility to avoid these biases."

Joy Buolamwini's Algorithmic Justice League is battling bias in AI and ML; they helped persuade Amazon, IBM, and Microsoft to put a hold on facial recognition. Another interesting initiative is Responsible AI Licenses (RAIL), which restrict AI and ML from being used in harmful applications.

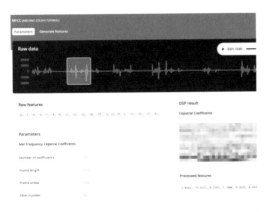

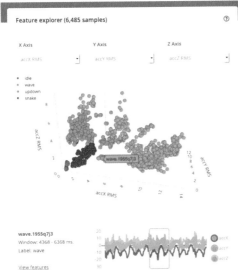

Edge Impulse lets you train ML models online for your microcontrollers, visualize sensor data, and more. It supports Arduino, OpenMV, and other popular TinyML platforms. (Top) Audio cough detection on an Arduino Nano 33 BLE Sense, and (above) multi-axis accelerometer data visualized in 3D. edgeimpulse.com

Shawn Hymel built this machine-learning Lego Finder by training a model on Edge Impulse, then running TensorFlow Lite on the OpenMV H7 Camera microcontroller. openmv.io/products/openmv-cam-h7

MAKER-FRIENDLY MACHINE LEARNING

There's also a lot of work to do to make ML more accessible to makers and developers, including creating tools and documentation. Edge Impulse is the exciting new tool in this space, an online development platform that allows makers to easily create their own ML models without needing to understand the complicated details of libraries such as TensorFlow or PyTorch. Edge Impulse lets you use pre-built datasets, train new models in the cloud, explore your data using some very shiny visualization tools, and collect data using your mobile phone.

When choosing a board to use for your ML experiments, be wary of any marketing hype that claims a board has special AI capabilities. Outside of true AI-optimized hardware such as Google TPU or Nvidia Jetson boards, almost all microcontrollers are capable of running tiny ML and AI algorithms, it's just a question of memory and processor speed. I've seen the "AI" moniker applied to boards with an ARM Cortex-M4, but really any Cortex-M4 board can run TinyML well.

Shawn Hymel is my go-to expert for machine learning on microcontrollers. His accessible and fun YouTube videos got me going with my own experiments. I asked him what's on his checklist for an ML microcontroller: "I like to have at least a 32-bit processor running at 80MHz with 50kB of RAM and 100kB of flash to start doing anything useful with machine learning," Hymel told me. "The specs are obviously negotiable, depending on what you need to do: accelerometer anomaly detection requires less processing power than voice recognition, which in turn requires less processing power than vision object detection."

The two ML-ready boards I've had the most fun with are Adafruit's EdgeBadge and the Arduino Nano 33 BLE Sense. The EdgeBadge is a credit card-sized badge that supports TensorFlow Lite. It has all the bells and whistles you could hope for: an onboard microphone, a color TFT display, an accelerometer, a light sensor, a buzzer and, of course, NeoPixel blinkenlights.

Here I'll take you through the basics of how to sense gesture with an Arduino Nano 33 BLE Sense using TinyML. I also recommend Andrew Ng's videos on Coursera and Shawn Hymel's

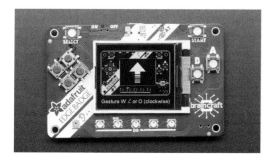

Adafruit's EdgeBadge runs TensorFlow Lite on an SAMD51 Cortex-M4 MCU; it's essentially an upgraded version of the PyBadge with a new onboard microphone. Their PyGamer board can run the same ML software on the same chip. adafruit.com/product/4400 and /4242

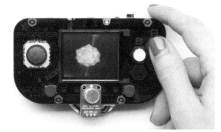

Arduino's Nano 33 BLE Sense board crams a Cortex-M4 MCU and tons of sensors to support machine learning into a tiny, wearable form factor. store.arduino.cc/usa/nano-33-ble-sense

excellent YouTube series on machine learning (youtube.com/user/ShawnHymel). You can take a TinyML course on EdX (programs.edx.org/harvard-tiny-ml) that features Pete Warden from Google, who coined the term and also authored a fantastic book on the subject. For keeping up-to-date on the latest developments, seek out Alasdair Allan's interesting and informative articles and blog posts on embedded ML; you can find him on Twitter and Medium @aallan.

TINY ML IN THE WILD:
MACHINE LEARNING PROJECTS FROM THE COMMUNITY

CORVID-19: CROW PAPARAZZO
by Stephanie Nemeth

The longing for interaction during the Covid-19 lockdown inspired Stephanie Nemeth, a software engineer at Github, to create a project that would capture images of the friendly crow that regularly visited her window and then share them with the world. Her project uses a Raspberry Pi 4, a PIR sensor, a Pi camera, Node.js and TensorFlow.js. She used Google's Teachable Machine to train an image classification model in the browser on photos of the crow. In the beginning, she had to continually train the model with the newly captured photos so it could recognize the crow. A PIR sensor detects motion and triggers the Pi camera. The resulting photos are then run through the trained model, and, if the friendly corvid is recognized, the photos are tweeted to the crow's account using the Twitter API. Find the crow on Twitter @orvillethecrow or find Stephanie @stephaniecodes.

TINY ML IN THE WILD

NINTENDO VOICE HACK by Shawn Hymel

Embedded engineer and content creator Shawn Hymel imagined a new style of video game controller that requires players to yell directions or names of special moves. He modified a Super Nintendo (SNES Classic Edition) controller to respond to the famous *Street Fighter II* phrase *hadouken*. As a proof-of-concept, he trained a neural network to recognize the spoken word "hadouken!" Then he loaded the trained model onto an Adafruit Feather STM32F405 Express, which uses TensorFlow Lite for Microcontrollers to listen for the keyword via MEMS microphone. Upon hearing the keyword, the controller automatically presses its buttons in the pattern necessary to perform the move in the video game, rewarding the player with a bright ball of energy without the need to remember the exact button combination. Shawn has a highly entertaining series of videos explaining how he made this project — and other machine learning projects — on Digi-Key's YouTube channel youtube.com/digikey. You can also find him on Twitter @ShawnHymel.

PROJECT:
RECOGNIZE GESTURES USING MACHINE LEARNING ON AN ARDUINO

Since I got my hands on the Arduino Nano 33 BLE Sense at Maker Faire Rome last year, this little board has fast become one of my favorite Arduino options (Figure Ⓐ). It uses a high-performance Cortex-M4 microcontroller — great for TinyML — and loads of onboard sensors including motion, color, proximity, and a microphone. The cute little Nano form factor means it works really well for compact or wearable projects too. Let's take a look at how to use this board with TinyML to recognize gestures.

1. SET UP THE BOARD
As usual, you'll start off by installing the board on your Arduino IDE. Go to Tools→Board→Board Manager and search for Nano 33 BLE. Install the package labelled Arduino nRF528x Boards. Next you need to install two libraries. Go to Tools→Manage Libraries and install Arduino_TensorFlowLite and Arduino_LSM9DS1.

Connect your Arduino Nano 33 BLE Sense to your computer using a micro USB cable, then go to Tools→Board and select Arduino Nano 33 BLE. Next, go to Tools→Port and make sure your board is selected there too.

2. CHECK OUT THE SENSORS
For our mini ML project we're going to be sensing gestures, so let's run some example code to try out the motion sensors on our board. Earlier you installed a library for the LSM9DS1, which is a motion sensor module with nine degrees of freedom (9-DOF) made up of a 3-axis accelerometer, a 3-axis gyroscope, and a 3-axis magnetometer. This means it can detect three key aspects of movement:

Shawn Hymel, www.kittyyeung.com

acceleration, angular velocity, and heading. Go to File→Examples→Arduino_LSM9DS1 and choose one of the three example sketches to try out. Upload it to your board using Sketch→Upload. Take a look at the values from your motion sensor using Tools→Serial Plotter or Serial Monitor, then give your board a good wiggle to watch the values change. Here's what my gyroscope test data looked like on the Serial Plotter (Figure **B**).

If you're new to 9-axis sensors, spend some time playing with the three example sketches to get comfortable with how the accelerometer, gyroscope, and magnetometer work. Each sensor gives us lots of valuable information on its own, but to get the most accurate picture of what our board is up to we're going to use two sets of sensor readings. Also, because we're going to use this data in a machine learning project for training a model, we need to be able to capture those readings in the right format. Luckily for us, the lovely people at Arduino have already written some code to help us do just that. Let's take a look at their example code (Figure **C**, next page).

This code reads the data from the accelerometer and the gyroscope and prints the values in a CSV format with headers that will help train our model. Open up your web browser and navigate to Arduino's TensorFlow Lite tutorials on Github (tinyurl.com/y33da3qe) to find the code snippet *IMU_Capture.ino* that I'm using here. Send the code to your board using Sketch→Upload, then go to Tools→Serial Plotter.

3. CAPTURE SOME GESTURES

It's time to play with gestures! Take the board in your hand and try different gestures to see what the data looks like. For my first gesture, I punched my fist forward four times, which gave me a very satisfying visualization of the data coming from my motion sensors (Figure **D**). For my second gesture, I tried a *Karate Kid*-style "wax on, wax off" motion which gave me a less dramatic but still recognizable pattern (Figure **E**).

Spend some time to find a couple gestures you really enjoy. Let your imagination run wild! You can strap it to your leg if you fancy experimenting with machine learning and yoga poses, or onto your wrist if you're more into tennis, or even onto your head if you've ever wondered what it's like to

TINY ML IN THE WILD

CONSTELLATION DRESS by Kitty Yeung

Kitty Yeung is a physicist who works in quantum computing at Microsoft. She also makes beautiful science-inspired clothing and accessories, including a dress that uses ML to recognize her gestures and display corresponding star constellations. She uses a pattern-matching engine, an accelerometer, an Arduino 101, and LEDs arranged in configurations of four constellations: the Big Dipper (Ursa Major), Cassiopeia, Cygna, and Orion. Yeung trained the pattern-matching engine to memorize her gestures, detected by the accelerometer, to map to the four constellations. To learn about the project or see more of her handmade and 3D-printed clothing, check out her website kittyyeung. com or find her on Twitter @KittyArtPhysics.

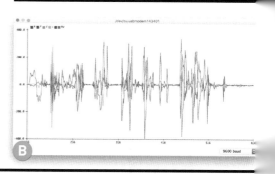

be a giant joystick (haven't we all?).

Once you've chosen your gestures, you can start gathering the data needed to train your model. Close the Serial Plotter, press the reset button on the board, and open up the Serial Monitor. Do your first gesture at least ten times, in my case punching. You should see a whole load of data print to your Serial Monitor (Figure **F**).

Copy all the data in the Serial Monitor into a plain text file and name it *punch.csv*. (You don't have to use a punch as your gesture here, but we're using a model that's ready-made for us by Arduino, which requires the filename *punch.csv*.) Clear the output on your Serial Monitor and press reset on the board again, then repeat the process for your second gesture. This time, call your file *flex.csv*. You're ready to train your model.

4. TRAIN YOUR MACHINE LEARNING MODEL

We're using Google Colab to train the machine learning model in a web browser. Head to the Tiny ML on Arduino gesture recognition tutorial (tinyurl.com/yxewlap2) to get started (Figure **G**).

On the Google Colab page you'll see a panel on the left-hand side containing icons and menus for Table of Contents, Code Snippets, and Files. In the main work area of the page you'll find a step-by-step tutorial with headers and code "cells" that you can run by pressing the Play icon.

Start off by running the Setup Python Environment cell, then upload *punch.csv* and *flex.csv* to the Files section. You can then run cells to graph your data and train your neural network, before building and training your model. Finally, run the cell called Encode the Model in an Arduino Header and download the resulting *model.h* file.

5. RUN YOUR MODEL ON THE ARDUINO

To run the model trained on your data, you need to add it into a sketch on the Arduino. In your web browser, navigate back to Github to download the *IMU_Classifier.ino* code I'm using here (tinyurl.com/y3kv3sul). Open up the sketch in your IDE (Figure **H**) then create a new tab by clicking the down arrow in the top right corner then selecting New Tab. Call this tab **model.h** and insert the contents of your downloaded *model.h* file.

Compile and run the sketch, then open the

Nathan Griffith

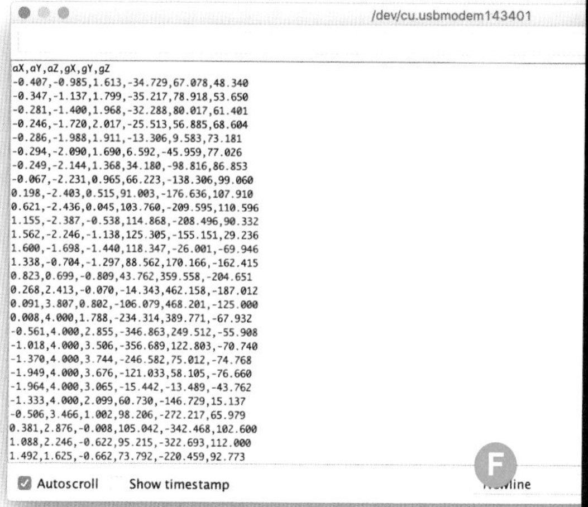

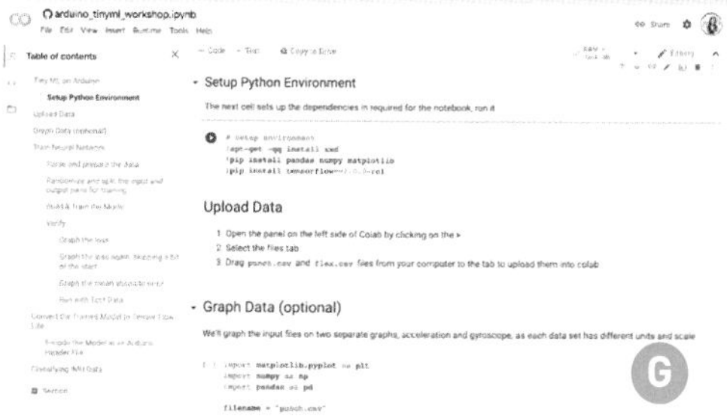

Serial Monitor (Figure I) and try out your gestures. You should see **punch** and **flex** followed by a number between 0 and 1. These numbers tell you what ***degree of confidence*** your model has in categorizing what you're doing as one or the other of your trained gestures, with 0 being low confidence and 1 being high confidence.

PUNCH ABOVE YOUR WEIGHT

If it's all working, you should feel very proud of yourself. You have managed to train and run a machine learning model on an Arduino board — an awesome new skill.

These outputs can be used for all sorts of fun applications. The Arduino team used their model to type emojis with gestures, but you can use any outputs you like: servomotors, blinky lights, noisy alarm systems, or sending secret signals to another device.

Now that you've got to grips with the basic process, you can also replicate this project with sets of data from different sensors. For your next Arduino and TinyML project I'd highly recommend experimenting with the tools from Edge Impulse to help you get creative with machine learning.

TINY ML IN THE WILD

WORM BOT by Nathan Griffith

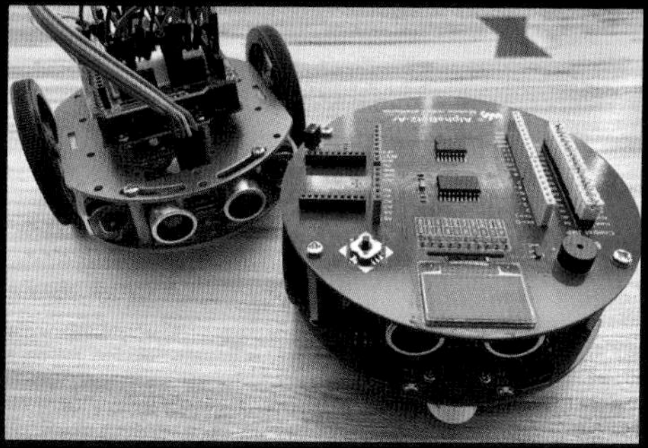

Artificial neural networks can be taught to navigate a variety of problems, but the Nematoduino robot takes a different approach: emulating nature. Using a small-footprint emulation of the *C. elegans nematode's* nervous system, this project aims to provide a framework for creating simple, organically derived "bump-and-turn" robots on a variety of low-cost Arduino-compatible boards. Nematoduino was created by astrophysicist and tinkerer Nathan Griffith, using research by the OpenWorm project, and an earlier Python-based implementation as a starting point. You can find a write up of his wriggly robot project on the Arduino Project Hub by searching for "nematoduino" and you can find Griffith on Twitter @ComradeRobot.

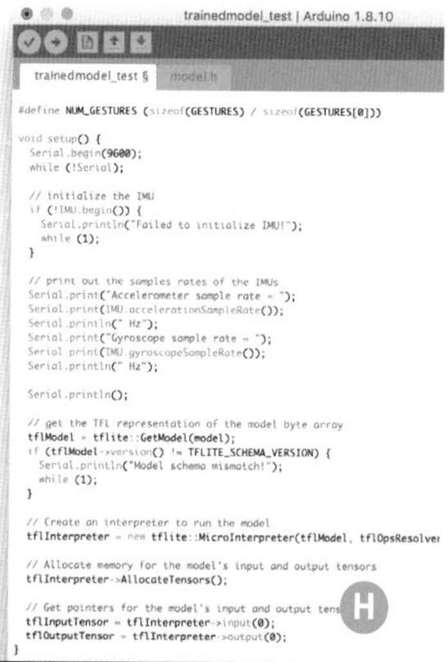

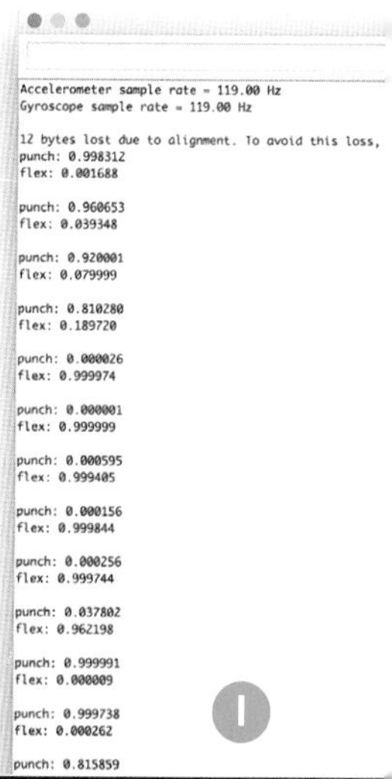

Smart Stop Signal

Use an Arduino and LED strip lights to dock your land-yacht with precision

Written and photographed by Krish Gupta

TIME REQUIRED:
1 Hour

DIFFICULTY:
Easy

COST:
$20

MATERIALS
» **Arduino Uno R3 microcontroller board** or similar. I used an Arduino Nano for the smaller, finished version.
» **Ultrasonic sensor, HC-SRO4**
» **Solderless breadboard**
» **LED strip lights, WS2812B (30 LEDs)**
» **Jumper wires (10)**
» **DC power source, 5V 500mA**
» **Pushbutton, momentary (optional)**

TOOLS
» **Computer with Arduino IDE software** free at arduino.cc/downloads
» **Soldering iron and solder**
» **Multimeter**
» **Pliers and basic home tools**

We live in a house with a very tight garage, and our relatively big car leaves a tiny space in front. I designed and built this simple project to visually guide the driver into the optimal parking distance, using real-time feedback about how close the car is to the wall in front. In short, it tells you when to stop.

The project uses an ultrasonic sensor and Arduino to measure distance and then displays it on a full-color LED strip as a progress bar. The sensor is mounted on the wall; as the distance between the car and wall shrinks, the light strip shows an increasing number of illuminated LEDs, which also change color from green to amber to red, and then finally, flashing red.

This quick build will help anyone safely pull their car into the garage without worrying about hitting the wall or damaging the car. It is easy to assemble, and it's a fun project for learning how how to work with LED strips and how to use ultrasonic sensors to measure distances.

1. CONNECT THE LED STRIP

The green wire from the LED strip is the Data wire, red is Power, and white is Ground. Connect the white wire to the ground of your power source, and the red wire to 5V of the power source. Connect the green wire to Digital 3 on the Arduino (Figure **A**). I used a 1 meter, 60 count LED strip and cut that in half to make two separate garage sensors. You can use whatever number of LEDs works for your garage, just make sure you change the **NUM_LEDS** value in the Arduino code to whatever number you have.

2. CONNECT THE ULTRASONIC SENSOR

Place the ultrasonic distance sensor on the breadboard. Then connect it to the Arduino: sensor pin Trig to Arduino pin Digital 10, Echo to Digital 11, GND to GND, and Vcc to 5V (Figure **B**).

3. PROGRAM THE ARDUINO

Download the project code for free from makezine.com/go/parkingsensor and upload it to your Arduino (Figure **C**). That's it.

CRASH NO MORE

Now that you have your finished product you can use it in your garage!

Mount the ultrasonic sensor and LED strip on the wall and start bringing your car into the garage. Watch the colors change, and when the red lights flash, stop! The default stop distance is 30cm (about 1ft) but you can change that in the **ZONE** and **MAXRR** variables in the code.

If you want to mount this without the breadboard, you can use an Arduino Nano and put that inside a 3D printed case (the template I made is available in the code downloads folder, or you can modify it from tinkercad.com/things/5cLhsWooe17) but I just used a plastic jar lid I had lying around.

I'm also working on an advanced version of the code that lets you add a momentary pushbutton switch to go into a calibration mode where you can read and calibrate parking distances from the sensor itself and not have to hardcode it. You can find the code for this on the download link as well: makezine.com/go/parkingsensor.

Happy parking! ●

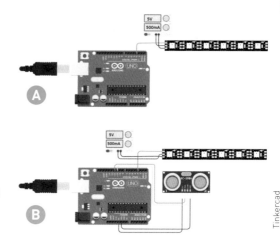

Tinkercad

```
Garage_Parking_Sensor_-_Basic
#include <FastLED.h>

#define trigPin        10     // ultrasonic sensro trig ping
#define echoPin        11     // ultrasonic sensro echo ping
#define NUM_LEDS       60     // How many leds in your strip?
#define DATA_PIN       3      // data pin for LED strip
#define ZONE           30     // default zone distance is 30 cms / a foot
#define MAXRR          30     // default distance for Flashing Red is 30 cms / a foot
#define MAXR           MAXRR+ZONE   // distance for Red
#define MAXY           MAXR+ZONE    // distance for Yellow/Amber
#define MAXG           MAXY+ZONE    // distance for Green
#define MAXFLASHCOUNT 30      // max times to flash

CRGB leds[NUM_LEDS];          // Define the array of leds

float duration, distance;
int flashcount, LEDpercm;

void setup()
{
  // initialize the LED strip
  FastLED.addLeds<WS2812B, DATA_PIN, GRB>(leds, NUM_LEDS);  // GRB ordering is typical
  FastLED.clear();  // clear all pixel data
  FastLED.show();

  // set the trig and echo pin modes
  pinMode(trigPin, OUTPUT);
  pinMode(echoPin, INPUT);

  LEDpercm = NUM_LEDS / ZONE ; // number of LEDs to glow per cm of distance measured

}

void loop()
{
  CallSensor();
}
```

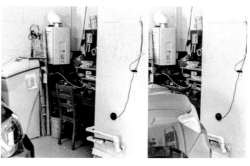

KRISH GUPTA is 13 years old and inventor of the Garage Parking Sensor. He picked up Arduino during Covid-19 and is still learning it. In his free time he likes to play soccer, drums, and guitar.

Hot Metal

Written and photographed by Howard Sandroff

I needed a blacksmithing forge — so I built one to my exact specifications

HOWARD SANDROFF
spent 45 years as an internationally recognized composer and music professor, after which he turned his attention to sculpting in steel. His artwork was profiled in vol. 74 of *Make:*.

Last year I decided I needed a forge so that I could advance my blacksmithing skills as another technique for my sculpting. There are many commercial models available but, being retired, I have more time than money so I decided to try my hand at making one.

To start, I studied YouTube videos of DIY forges. There are many and I watched them all. None were exactly what I wanted so I set out to design a hybrid. I researched the necessary materials. I read all about soft firebricks, which can withstand temperatures above 1,800°F — good for heating up mild steel. I'd need a high-temperature mortar, called "Satanite," of all things. Long hose clamps to hold things together and some steel angle iron for the frame. And lastly, the burner, which is challenging to make.

Adobe Stock - artinspiring

TIME REQUIRED:
2 Weeks

DIFFICULTY:
Advanced

COST:
$325

MATERIALS

» **Firebricks, soft insulating, 9"×4½"×2½" (26)** rated to 2,600°F
» **Satanite refractory mortar, 5 lbs**
» **Hose clamps** to go around the forge (~50") twice. Attach 4 or 5 big clamps together to go around once.
» **Angle iron, 1½"** I used a discarded steel bed frame.
» **Steel nipple, pipe or round tube, ³/₁₆"×3"×6"**
» **Steel plate, ³/₁₆"×4"×6¼"**
» **Steel strips, 1"×9¼"×⅛" (12)**
» **Hex bolts, ⅜"×16"×1½" (3)**
» **Quick disconnect hose coupling** for propane hose
» **Forge burner** I bought mine on etsy.com.
» **Spray paint, black high heat (1 can)**
» **Masking tape, 2" (1 roll)**

TOOLS

» **Screwdrivers, large and medium**
» **Pliers, medium, long nose and electrician's**
» **Hammer** Claw is fine. Ball peen is better.
» **Locking pliers, medium** aka vise-grips
» **File, 8" medium bastard**
» **Box cutter with replaceable blade**
» **Tape measure, at least 60"**
» **Ruler/straightedge, 12"**
» **Adjustable wrench** or set of open/box wrenches
» **Electric drill, bits, sanding disc, wire brush**
» **Drill bit, ⁵/₁₆"**
» **Tap and handle, ⅜"×18"**
» **Table saw, or hand saw and miter box**
» **Welder: MIG, TIG, stick, or gas**
» **Eye protection, N95 respirator, leather gloves**

CUTTING FIREBRICKS

The bricks are sized 4½"×2½"×9" so I decided that assembling them in the shape of a nonagon would give me the inside dimensions suitable for the kind of work I do. I needed to compute the angle to cut the bricks so that I could make a 10"-diameter forge as deep as a single course of bricks, 9". The bricks are soft enough to cut with a hand saw, but since I own a table saw I could set the blade at the correct angle and the fence to the correct width to cut them consistently.

Firebricks are pricey, and therefore I decided to first model the cuts using some short 2×4s. I ran each 2×4 through the saw twice, once for each side. Then I arranged them in the proper shape, sawed-corner to sawed-corner and voilà, created a nonagon (Figure Ⓐ).

Using the same technique, I cut all nine bricks to give the long edges a 20° slant. I used a worn-out plywood blade (and scratched up the table) and the process worked — I could now build my nonagon forge with an inside diameter of 10".

Reading the instructions for using the Satanite, I learned to mix only a small amount at a time, as the mortar dries very quickly. I smeared the mortar between each pair of bricks to cement them together. Once I finished the entire nonagon, I placed a long hose clamp on the top and bottom to hold it tightly as it dried overnight.

Next step was to coat the interior of the forge with multiple layers of the Satanite (Figure Ⓑ). These layers protect the bricks from the heat and can be reapplied when they're burned, thereby prolonging the life of the firebricks.

Ⓐ

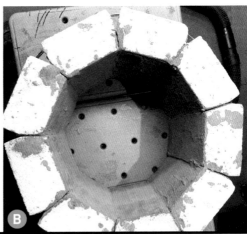
Ⓑ

C

WELDING THE BRACING

Now I needed to build a steel frame to hold the bricks together as the mortar didn't quite do the job. One of my scrounging habits is to occasionally cruise the neighborhood looking for cast-away bed frames. I have a great collection, and as the bed frames are usually made of 1½" angle iron, they were perfect for the forge frame. Using a steel-cutting chop saw, I cut all the pieces at the correct angle to place at the front and back of the forge. I spot-welded them together as I went, and when the two ends were complete, I welded some strips of steel along the length of

the forge to connect the end frames, and a steel plate on top to make a stand for the burner. I then masked off all the bricks and painted the frame and stand a high-temperature black (Figure C).

MOUNTING THE BURNER

After reading and viewing many articles and videos about making a burner, I decided that my tools and skills were not up to the precision modifications of the necessary brass and steel fittings. Fortunately, I found an affordable burner on Etsy. Once that arrived, I drilled the required hole through the forge and the steel plate using a

3" hole saw, then mounted a short length of pipe, modified to hold the burner (Figure 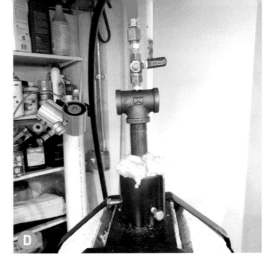). I drilled and tapped three ⅜" holes spaced equally around the perimeter of this pipe, at a height where they're best positioned for my burner. The pipe is welded to the steel plate to hold it in place; the burner is held in place with three ⅜" bolts.

Now I was ready to "cure" the forge. I heated it up for an hour each day for a week, progressively increasing the heat so that any residual moisture in the bricks boiled away.

HAMMER TIME

Finally I was ready to try the forge with a piece of steel. Using the balance of the firebricks to close the front and back of the forge, leaving a small opening for the metal to be heated and airflow for combustion, I heated up my first rod of steel and proceeded to beat it into a flat shape using a hammer and my anvil. Success! I now had a working forge. To finish the project, I built a roll-around stand for the forge and propane tank out of the same bed-frame angle iron (Figure ). Total cost, including the burner: about $325.

Building this has been very gratifying. Although I am by no stretch of the imagination a blacksmith, I do use their techniques in my sculpting. I use the forge at least once a week and it is holding up very well. I would, however, like to figure out how to better insulate it so that I can reach hotter temperatures.

I will also say, without hesitation, that the forge should *absolutely not* be used for roasting marshmallows. Marshmallows are full of sugar and are bad for your health. Besides, if you stick a marshmallow in the forge it will vaporize in a millisecond. ✪

My sculpture *Three Sisters*. Hammered, welded steel with oxidized patina.

Cutting Edge Contributions

Makers everywhere are sharing great builds on makeprojects.com — post yours too!

RENZONICA
Bryan Salt
@ModernMaestro
Five years in the making, this unique 19-note, semi-automatic instrument uses servomotors and standard Hohner harmonica reed plates to play pre-selected music "rolls" like a Pianola, except it's not fully automated — the musician blows or draws into the instrument to create each note, then releases a button to automatically select the next correct note.

makeprojects. com/project/the-renzonica-a-musical-instrument-anyone-can-play

A

makeprojects.com/project/making-a-stringless-tin-can-telephone-with-arduino

B

makeprojects.com/project/wooden-wanderlust-map

Ⓐ STRINGLESS TIN CAN TELEPHONE

Geoff McIntyre @FacioErgoSum

A modern reboot of a classic children's toy, this Arduino-based 2.4G walkie-talkie provides a low barrier for entry to the world of hobby electronics. Yes it really works, up to 1 kilometer range!

Ⓑ WOODEN WANDERLUST MAP

Binh Thai @Binh

Adorn a wall or mirror with your world travels, using CNC to cut out wooden continents and 3D printing to make location markers.

Ⓒ 3-LEGGED WALKING ROBOT

Yutaka Sato @VoidPump

Could three-legged locomotion achieve the mobility of two-legged walking and the stability of four-legged walking? This unique tripod robot's got a funky, slide-winding gait that does not exist in the natural world!

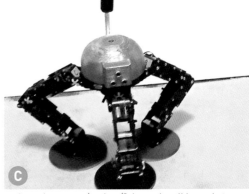

Ⓒ makeprojects.com/project/3-legged-walking-robot

Ⓓ MORSE CODE TRANSCRIBER

Jane Frauenfelder @BionicErebus

This thumb-sized handheld device lets you practice Morse code or just write things with your hand in your pocket without anyone noticing. It uses an M5Stick programmed in Arduino to convert button presses from Morse dah-dits to alphanumeric text. Upgrade it to wirelessly send messages to a computer or a friend with a similar device.

Ⓓ makeprojects.com/project/mini-morse-code-transcriber

Ⓔ DIY ELECTRIC PENNY SKATEBOARD

Marcus Lee Wei @catleemiaw

Singapore is a hot and humid concrete jungle. This electric pennyboard is perfect for moving around easily, belt driven by a 6254 motor, with longboard trucks and a tough 3D printed PETG enclosure.

Ⓔ makeprojects.com/project/lightweight-diy-electric-pennyboard-6kg-

Ⓕ TEXT TO BRAILLE

Efe Tascioglu @EfeTascioglu

A high schooler prototyped this gadget to make any book readable by the blind. She uses a Raspberry Pi camera to scan the pages and translate them to Braille, then an Arduino fires off the Braille patterns in LEDs — just replace with solenoids to create a readable tactile display! ⊘

Ⓕ makeprojects.com/project/text-to-braille

META TIPS FOR MAKERS

When your frame of mind is as important as the tools you use

Written by Gareth Branwyn

Sometimes, the most potent ideas that impact your work-life are not practical shop practices, but higher-order thoughts that organize your thinking and inform your approach to making. Here are six such "meta" tips.

MAKE IT SEEM HARD, NOT BE HARD

When teaching someone something new, reward them with success as early as possible. A great example is the light-up Learn to Solder badge at Maker Faire. Let them get their hands dirty ASAP with a simple project, so they can exercise their newfound skills and be rewarded for their efforts. Make it seem hard, not be hard.

DETAILS LAYER

Scott Wadsworth of *The Essential Craftsman* has a wonderful statement that speaks to something very important in all aspects of making (and in life): details layer. In every step of a project the precision with which you do one step, one layer, carries over into the next, and the next, and the next. Over time, mistakes and imperfections compound, so it's important to do each step as well and as thoughtfully as possible.

THE KENNY ROGERS RULE

The extent to which you don't want to drop what you're doing and take a break ("I know I can fix this, damn it!") is inversely proportional to the extent to which you need to take that break. Why Kenny Rogers? Cause as the late Country & Western star so wisely told us: "You got to know when to hold 'em, know when to fold 'em, know when to walk away..."

IDEA FISHING WITH DAVID LYNCH

Filmmaker David Lynch says: "Ideas are like fish. If you get an idea that's thrilling to you, focus your attention on it and these other idea-fish will swim to it. It's like bait. They'll hook on to it and you'll get more ideas. And you just reel them all in."

KNOW WHERE YOUR TOOLS AND MATERIALS ARE GOING

As you work with a tool, be mindful about the direction of its rotation (and other movement), speed, point of contact with the material, possible points of failure, and where the tool and the material are likely to end up if something goes wrong. This will help keep you out of the path of failure and safer should something go wrong.

MAKING IT PERFECT ENOUGH

Andy Birkey, who restores Gothic churches, says that when you start a project you want everything to be perfect, but as the deadline draws closer, you need to start making an important pivot — letting go of absolute perfection. In the high ceilings of old churches, he says, you can see imperfections that previous builders left because they knew they'd never be seen from the ground. It doesn't need to be perfect. It just needs to be perfect enough.

This article is excerpted from *Tips and Tools from the Workshop, Vol. 2* by Gareth Branwyn, coming soon in 2021! Check out *Tips and Tales from the Workshop, Vol. 1*, available now from makershed.com and fine bookstores.

GARETH BRANWYN is former editorial director of *Make:*, a *Wired* alum, and a current contributor to *Boing Boing*, Adafruit, and *Hackspace*. He publishes the newsletter *Gareth's Tips, Tools, and Shop Tales* at getrevue.co/profile/garethbranwyn.

Victorian Toys and Flatland Rockets

Make a miraculous Marangoni soap boat

Written and illustrated by Bob Knetzger

A

B

C

Here's an update of a Victorian plaything. Cut out the fish on the dotted line and float it on a pan of water. Place a single drop of olive oil in the circle. The oil quickly spreads out the slit and across the water. The fish "swims" in the opposite direction, like an exhaust-spewing rocket subject to Newton's third law of motion. Sadly, the soggy paper fish is only good for just a single use (Figure A).

NOW TRY THIS NEW, MORE DURABLE VERSION:

Find a flexible lid from a margarine or yogurt container. Look for the recycling symbol 2 or 4 for low- or high-density polyethylene. (PE is one of the few plastics that floats.) Use a paper punch to make a small circular hole, then cut out the "rocket" shape as shown in Figure B.

Float the rocket in a pan of clean water. Dip the tip of a toothpick in detergent and momentarily touch it inside the rocket's round hole. As the detergent dissolves, it spreads down the slit and out along the surface of the water — the rocket shoots forward! Touch it again. After a time or two, you'll have to change the water for the effect to work again. Visit makezine.com/go/marangoni-boat to see a video demo.

Another force is also at work: the Marangoni effect, the difference in surface tensions created by the molecules of detergent as they make the water slipperier and "wetter." The surface tension is reduced behind the rocket, causing the water in front to contract, pulling the rocket forward (Figure C).

These tensions, forces, and actions all exist at the single-molecule-thick surface of the water — similar to the two-dimensional world in Edwin Abbott's Victorian-era book, *Flatland: A Romance of Many Dimensions.* ◗

BOB KNETZGER is a designer/inventor/musician whose award-winning toys have been featured on *The Tonight Show*, *Nightline*, and *Good Morning America*. He is the author of *Make: Fun!*, available at makershed.com and fine bookstores.

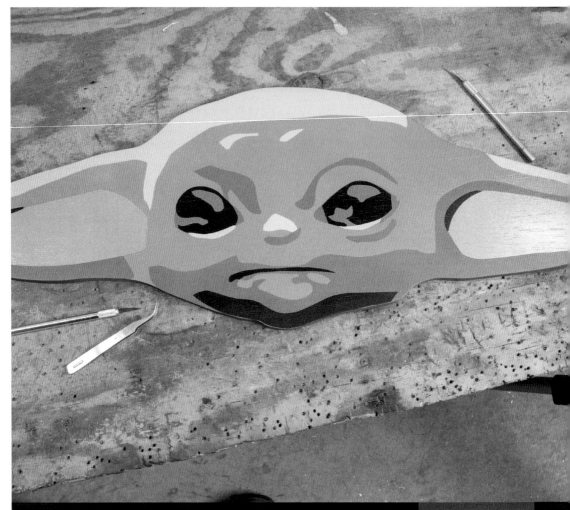

HAND-CUT PAINT MASKING

WESLEY TREAT is an author, maker, and content creator, whose projects he hopes will inspire others to try new things. You can find his videos at youtube. com/WesleyTreat.

How do you create a multi-layer, positive paint mask for artwork, without a CNC vinyl cutter? This is the way. **Written by Wesley Treat**

A number of low-cost vinyl-cutting machines have entered the market in recent years, enabling hobbyists to make their own adhesive-vinyl stickers, stencils, and paint masks at home. It's entirely possible, though, to create even a complex, multi-layer paint mask with little more than a roll of vinyl and a fresh X-Acto blade.

1. CHOOSE AN IMAGE TO PAINT

This can be nearly anything you want, as long as it can be broken down into a handful of colors and isolated shapes, much like a "paint by numbers" image. Clip art works well, or if you have more skill than I do, you can draw something by hand.

For this demonstration I've chosen everyone's favorite macrotous character from *The Mandalorian*. I downloaded an image online, isolated his head in Adobe Photoshop (Figure **A**) and used Adobe Illustrator's "Image Trace" feature to convert it to vectors (Figure **B**). After some experimentation with the settings, and a bit of manual editing, I was able to simplify the image into five distinct colors (Figure **C**). In the end, his innocent expression ended up looking more like stern disapproval, but I thought having that face hanging above my workbench might do well to keep me on task.

If you don't have access to Adobe's software, there are free options available for download, like GIMP and Inkscape. Just search online for "how to vectorize an image."

2. CREATE YOUR TEMPLATE

Draw your image at the actual size you intend to paint it; simple outlines are ideal. In my case, I converted my shapes' colors to black outlines (Figure **D**), saved the image as a PDF, and took it to my local copy shop to print it out on their large-format printer for a couple of bucks. Either way, make your template on paper no thicker than common printer paper. If you like, you can download my template at makezine.com/go/paint-masking.

3. PREPARE YOUR CANVAS

You can use just about anything you want as your canvas, as long as it's flat. For Baby Yoda — or the Child if you want to get all "actually" about it — I used my template to cut out a piece of birch

TIME REQUIRED
2–5 Days

DIFFICULTY
Intermediate

COST
$40–$60

MATERIALS
- Oramask 813 stencil film
- RTape Clear Choice transfer tape or similar
- Primer
- Paint just about any kind
- Substrate whatever you're going to paint on
- Artwork on paper Your can try my "Baby Yoda" template at makezine.com/go/paint-masking.

TOOLS
- Cutting mat
- Hobby knife X-Acto or similar
- Vinyl application squeegee (optional)
- Tweezers (optional)
- Dental pick (optional)
- Computer with vector graphics software (optional) You can free-hand your art, or use the computer (see step 1).

A

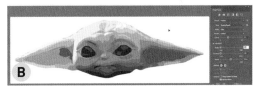

B

C

D

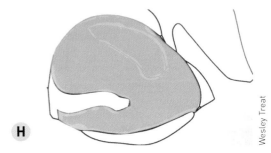

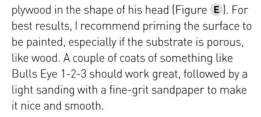

Wesley Treat

plywood in the shape of his head (Figure **E**). For best results, I recommend priming the surface to be painted, especially if the substrate is porous, like wood. A couple of coats of something like Bulls Eye 1-2-3 should work great, followed by a light sanding with a fine-grit sandpaper to make it nice and smooth.

4. PLAN YOUR ORDER OF OPERATIONS

There are two basic types of paint mask. The first is a *negative mask*, in which the shape to be painted is left uncovered and the mask protects everything else. For example, if you wanted to paint a number on a door, the shape of the number would be left as a window in the middle of the mask, and the mask would protect the surrounding area from the fresh paint. It's what you'd traditionally think of as a stencil.

The second is a *positive mask*. This is where you apply the paint first, then adhere the mask in the shape you want to preserve before applying the next paint color. This is the method we'll be using here, so we need to plan the order of our paint colors accordingly. I find it's easiest to start with the inner, isolated shapes, then move on to the larger, surrounding areas. In the case of Baby Yoda, I began with the black that made up his eyes and other small spots. The light green of his skin — the largest area — was saved till the end.

5. APPLY THE FIRST PAINT COLOR

You can use just about any type of paint. I've done vinyl masking with both oil-based and latex paints, and applied them with a spray gun, a brush, and a roller. Just be sure to follow the application instructions for the paint you choose. For this project, I went with rattle cans.

Apply a nice, even coat wherever the first color appears in the final image. Baby Yoda has black spots in multiple areas, so I sprayed the whole board to make sure I got thorough coverage everywhere I was going to need it (Figure **F**).

6. CUT THE FIRST MASKS

For most types of paint masking, I recommend Oramask 813 adhesive vinyl. It's semi-transparent, which makes it easy to line up with other elements. It comes in rolls of different widths and can be purchased from multiple vendors online.

With a cutting mat underneath, simply lay your paper template on top of the vinyl and hold it down with a couple of weights. A coffee mug or a tape dispenser will work — anything to keep the template from moving. Then, using a sharp, new blade, cut out your initial shapes using moderate pressure (Figure **G**). As you cut through the template, you want to perforate the vinyl without cutting the paper backing underneath. It shouldn't take much practice to get it right. In fact, you'll probably find that applying enough

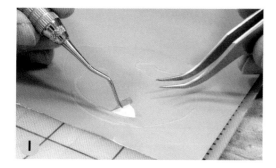

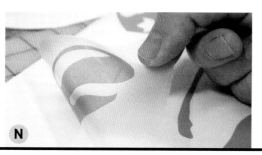

pressure just to cut through the paper template is about all you need to properly score the vinyl (Figure **H**).

7. DISCARD UNNEEDED VINYL

Once you've cut all the shapes for the first mask, you'll want to remove the superfluous material. This process is called *weeding*. The unneeded areas can be removed by picking up an edge with the point of your knife and peeling them away (Figure **I**). Tweezers, and sometimes a dental pick, can be helpful here, but aren't strictly necessary. With larger areas, I often find it easier to divide them up by scoring around my shapes and removing smaller sections at a time (Figure **J**), rather than wrangling wide sheets all at once. Just remember to think in reverse. You want to preserve the parts that protect your last painted color, and remove everything else.

8. TRANSFER YOUR MASK

For small, simple shapes, you can just peel up the mask and apply it to your canvas. To move unwieldy shapes, or multiple shapes at once, it's best to use what's called *transfer tape*. It comes in opaque and clear versions, and in rolls of varying widths, with either low or high tack (the amount of stickiness). For general use, I recommend a clear, high-tack selection like RTape Clear Choice AT65 (Figure **K**). There are similar, generic versions available as well.

Lay the transfer tape sticky side up and, starting at one end, carefully lay the sheet of weeded masking facedown onto the tape (Figure **L**). Then, flip it over and use a plastic squeegee tool to rub the transfer tape down and ensure adhesion (Figure **M**). You can also use something like a credit card or a plastic scraper. Start in the center and rub outward in all directions.

Finally, carefully peel up the transfer tape, lifting the vinyl mask away from the paper backing (Figure **N**).

9. APPLY THE MASK

Before applying your mask, be sure the previous layer of paint has set completely. Typically, I allow at least 24 hours before applying any mask. Otherwise, even if the paint is dry to the touch, you may have problems with paint peeling up when you remove the mask.

To apply the mask, simply lay it onto your canvas, starting along one edge of the transfer tape and rubbing it down as you go. Again, use the squeegee to ensure adhesion (Figure **O**). Then peel up the transfer tape slowly, leaving the vinyl masking in place. If some of the masking doesn't want to stay down, simply back up, rub it again with the squeegee, and try again. It helps to peel the transfer tape parallel to the canvas, rather than straight up (Figure **P**).

TIP: Here's a painter's tip to help prevent paint from bleeding under the edge of your mask: Before spraying the next color, apply a light coat of the previous color — the one underneath the mask (Figure **Q**). This helps to seal the edges, and any bleeding that occurs at this stage will be the same color as what's underneath. This isn't always necessary, but it can be helpful if your surface isn't super smooth.

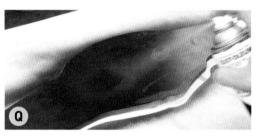

10. APPLY THE NEXT PAINT COLOR

Following the recoat times specified by your brand of paint, apply the next color. In Baby Yoda's case, I went with the pink of his ears (Figure **R**).

11. MAINTAIN YOUR BEARINGS

With the first mask, I was able to use Baby Yoda's adorable chin as a reference edge to align things. When dealing with isolated shapes, however, like the inside of the little guy's ears, you may not have anything to align to. This is where your paper template can serve another purpose. Place it directly on your canvas and use it to make a couple of tick marks with a pencil (Figure **S**). Use these marks to place your mask (Figure **T**).

To stay oriented, keep a copy of the final, color image nearby, so you can identify which shapes are which. It may also help to label your masks with a marker as you cut them (Figure **U**). Especially large or long shapes can be split in two to make them easier to handle and to make the most of your material.

12. CONTINUE WITH SUBSEQUENT LAYERS

Once you've completed the isolated shapes, move on to the wider areas. After I painted

Wesley Treat

U

Right Ear

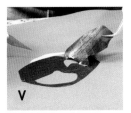

V

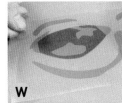

W

X

Y

Z

Baby Yoda's ears and the highlights on his head and nose, I addressed the dark green shapes. Since these areas encompassed regions I had already painted, I cut these masks to include the previous shapes and stacked them over the existing masks. On his right ear, for example, I cut a single, large piece that covered the pink, the black, and the dark green, and simply layered that on top. With his eyes, however, I opted to remove the previous mask first so I could see where to align things (Figure **V**). I just made sure to cut the new mask both to protect the existing black and to mask the new, dark green areas (Figure **W**). I finished with a final, even layer of light green paint (Figure **X**). There's no one, correct way to approach this! Just try to think ahead and visualize how you want the final image to look.

13. REVEAL YOUR IMAGE

Now comes the fun part! Once your last layer of paint has dried, you can remove the masks. Just like before, use the tip of your knife to pick up the vinyl's edge and carefully peel it away (Figures **Y** and **Z**). Layer by layer, your image will start to appear before your eyes. To me, it feels like a magic trick every time. I never get tired of it.

KEEPING IT POSITIVE

You might be asking why we didn't use negative masks instead. That's certainly an option, but it consumes a lot more material. To spray-paint even one small shape, you have to cover the entire canvas except for that area. Then, to paint subsequent shapes, you have to remove the entire mask and apply a new, full-size mask for each color. You'd save a little paint, but use a lot more vinyl. However, if you're brushing paint rather than spraying, you wouldn't have to worry about overspray. So, you could just apply negative masks using smaller rectangles of vinyl. Again, there's no one, correct way to do it. Choose the option that works best for you.

Now that you've completed your first image, try taking it a step further. Attempt a design with more shapes and colors. You can even add some visual texture or shading by lightly misting one color over another. Just plan ahead, be willing to make mistakes, and most important, have fun!

I have spoken. ✪

BESKAR INGOTS & WITCHER COINS

An introduction to pewter casting in laser-cut billet molds
Written and photographed by Dave Dalton

DAVE DALTON is the owner and operator of Hammerspace, Kansas City's makerspace. He has been making and breaking things ever since his grandfather showed him how to put a steam engine in a Ford Pinto, and has been metal casting and smithing since his teens.

TIME REQUIRED
An Afternoon

DIFFICULTY
Easy to Intermediate

COST
$20–$80

MATERIALS
» **Pewter, new or used.**
» **Cardboard**
» **MDF (medium density fiberboard)**
» **Baby powder**

TOOLS
» **Laser cutter** For the Beskar ingots, download our cutting files at makezine.com/go/beskar.
» **Electric melting pot (optional)** We've also used a hotplate or burner with an iron skillet or steel pot.
» **Spring clamps**
» **Flush cutters**
» **Leather work gloves, eye protection**
» **Hacksaw or band saw** for cutting Beskar sprue
» **Steel wool or scouring pads**
» **Dry-erase markers, black**

A

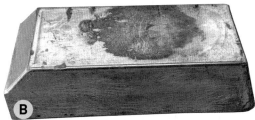

B

In preparation for Comic-Con cosplay season every year, our makerspace has an enormous amount of fun casting props from my favorite castable metal alloy: pewter. Pewter is a great metal for costume jewelry and props. It is nontoxic, easy to work, attractive, inexpensive, melts at a low temperature, and has a satisfying heft that makes a prop believable.

Pewter is an alloy consisting mostly of tin and a little antimony and copper. You can get pewter from a variety of sources. One of our favorites: dented pewter mugs and pitchers from the thrift store (Figure **A**). Ingots of pewter (Figure **B**) can be purchased online for about $20 per pound.

Pewter has a low melting point, 170°–230°C (338°–446°F), depending on the exact mixture of metals. You can melt it with a variety of heat sources. A bunsen burner, stand, and very small cast iron skillet work well. We have also used a hotplate and stainless steel pot for melting larger junk. We recommend that you use a small electric melting pot meant for lead and pewter (Figure **C**). These cost about $50 new and are worth every penny. Watching solid drinkware melt into a shiny silver puddle before your very eyes is mesmerizing — always a treat if it's your first time melting and pouring metal.

C

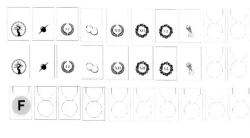

MAKING BILLET MOLDS

Once you have pewter, and a way to melt it, you'll need a mold. We use a process that allows us to make many types of molds for pewter using only our laser cutters and MDF or cardboard. Since paper doesn't combust below 451° Fahrenheit (thanks Ray Bradbury) then by extension thicker cardstock and MDF board can easily withstand 338°–446°F of molten pewter.

It is relatively easy to create the molds even without a laser cutter, using common hand tools like a scroll saw, scissors, and white glue. The ornament in the very center of Figure **D** was created in just such a handmade mold.

This process is very easy, inexpensive, forgiving of failure, and can be done safely with a few basic precautions and no special equipment. Our goal is to create an easy-to-assemble mold out of 3 or more layers of MDF and/or cardstock. This is called a **billet mold** (Figure **E**).

Figure D shows a variety of objects cast from pewter in billet molds. Most of these began as a drawing in software. In Corel Draw, I used vector lines for the profiles, and vector or raster images for raised textures and embossed areas. Some simple coins designs are shown in Figure **F** .

The fine lines will be cut by the laser, and the solid regions will be etched about 1mm into the surface of the MDF. To understand what these pieces do, let's skip ahead and look at the finished pieces of a mold.

CASTING WITCHER COINS

Each mold is made of a sandwich, or **billet,** of 3 or more layers. Figure **G** shows a simple coin mold. In this case, 5 layers are used: (from left to right) a blank front plate, the front etched surface plate with a pouring cup, the welting with pouring cup and downsprue, the back etched surface plate with pouring cup, and a blank back plate.

Let's look at these layers from the inside out. The center layer is called a **welting**. It provides the profile of the cast piece and also determines its thickness. In this case, we wanted a very thin coin, so we used cardstock (Figure **H**).

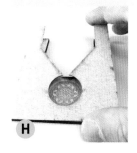

On each side of the welting is an **etched plate** (Figure **I**) that has been etched with a design and cut to match the profile of the welting, so that registration between the layers is easy. The dark areas will become raised surfaces on the finished coin. These can have a lot of detail or be simple and bold. Either way, molds with shallow designs survive the casting process well, but the deeper the etch needs to be, the shorter its lifespan becomes, as details may become trapped and be torn from the mold. I recommend dusting these pieces on new molds with baby powder before the first use. This will help trapped air to escape around the flowing metal in the tiny spaces between the powder particles and will vastly improve surface finish.

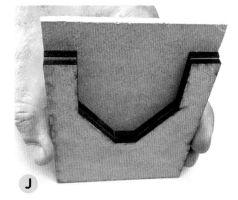

Figure **J** shows what the stack will look like once it is assembled between the blank **outer plates**, with only the top layer removed for viewing. The empty space at the top of this mold is very important and serves several purposes. It acts as a **pouring cup** to receive the molten pewter and as a **downsprue** to direct the pewter into the mold and allow air to escape. The relatively large pouring cup provides weight to generate pressure on the pewter below it, forcing it into the small details of the mold. In this case the mass of the pewter in the **button** (the leftover metal that cools in the pouring cup) is 30 times what is in the finished coin.

Once the details have been powdered and the billet stacked and aligned, we add spring clamps to hold the billet together and to make it stable and easy to pour into (Figure **K**). I like to make one of the outer plates taller than the rest of the mold so that I can pour against it as a backsplash.

I fill the molds to the top, in one swift pour (Figure **L** on the following page). This ensures that there's plenty of weight on the lowest details before they have a chance to cool.

CAUTION: Before pouring the molten pewter, don your work gloves and eye protection. The mold may steam or even smoke a bit the first time you use it. There are all kinds of chemicals that evaporate out of the cardboard and MDF during the first pour. Do this in a well-ventilated area; outdoors is best.

Once it's full, let your mold sit and cool for at least 15 minutes. There's no rush. You want the mold to be cool enough to handle, and the pewter inside can burn you if you're impatient.

Remove the clamps and begin to carefully pop the layers apart. It's really important to resist the urge to twist them like an Oreo cookie. This will tear the detail from your molds and create a lot of

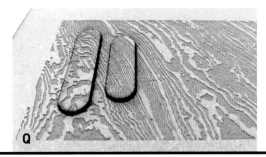

cleanup on the cast part. Just start at the top and gently separate the layers (Figure M). Then you can carefully clip the sprue with a pair of flush cutters, return the button to your melting pot, reassemble the mold, and cast another coin.

Finishing the pewter is easy. I like an antique look. The easiest way to achieve this is to grab a black dry-erase marker and give the entire coin a coat. Let the marker dry a bit and then give the coin a rubdown with some very fine steel wool. I suggest you use 0000 and stop when the luster and age look right to your eye (Figure N).

Congratulations, you now have coins you can toss to your Witcher or risk in a game of Gwent.

CASTING BESKAR INGOTS

Or perhaps you're a ruthless bounty hunter with a jetpack who prefers Beskar ingots to crowns and florins? Fear not, we've got you covered like IG-11. Here's how you can fill a camtono with the down payment on your black-market Baby Yoda.

Now, this prop involves a few new concepts, so we'll be building on what we've just learned. Billet molds are great for making flat shapes with a single layer of *embossed* design on the surface. But what if you need to *deboss* or indent the surface, like the Imperial stamp and oval detents in the front and back of this Beskar ingot (Figure O)? Those areas must be built up in your mold, rather than etched away. To solve this problem, we used the preferred choice of bounty hunters everywhere: more laser.

Much like our coin mold, we begin with the familiar layers: outer plates, etched plates, and welting (Figure P).

Figure Q is a close-up of the Damascus-like texture of the prop, etched into the MDF across the entire surface. Two ovals are cut from the back and a circle from the front, to allow those areas to be removed. The loose parts can then be adjusted, swapped out, and repaired if they suddenly initiate self-destruct, as is often the case with this challenging mold. Glue the oval parts back into place, at their new raised height, from the backside with wood glue to prevent leaking. Repeat this process to create the Imperial stamp on the front (Figure R).

The welting of the Beskar ingot hides its own clever tricks. Naturally, the Damascus-like

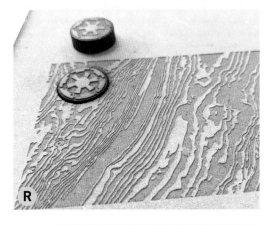

R

S

T

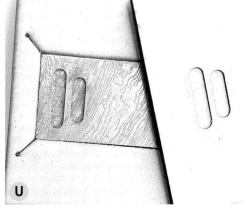

U

V

BESKAR INGOT VER. 2.1
BY HAMMERSPACE WORKSHOP
https://hammerspaceworkshop.com/

texture should also be visible at the ingot's edge (Figure **S**). In order to achieve this magic trick on a surface we cannot etch on, we instead made precise and tiny convolutions in the outline of the welting (Figure **T**) to create stripes of high and low that match the mirrored pattern of the faces. This makes it look like the full pattern is visible on 5 of the 6 surfaces of this prop. (If you're ambitious you can sculpt the top with a needle file after cutting away the sprue and button.)

This trick does not come for free. It makes it important to add some features to the welting that were not important before. First, it's now possible to flip this welting over and mess up the pattern, so registration lines were added on the left side of each piece. Second, the cast part is often locked into the welting by these irregular notches. This problem can be solved by adding some relief cuts to the welting as shown in Figure **U** . These let the welting stretch enough to release the ingot.

During demolding, it's not uncommon to damage one or more of the small debossed pieces. I always cut some extras so I can fix the mold. Other than that, Beskar ingots are just like coin molds, but they use a lot more pewter — at least a full pound of pewter to cast a 12oz ingot.

YOU CUT, WE CUT

We are pleased to share our drawings for free, so you can laser-cut your own Beskar ingot molds (Figure **V**). Download the files at makezine.com/go/beskar, and if you share them, please credit the team at Hammerspace Workshop.

Or, if your local makerspace or library doesn't have a laser cutter, we are able to cut and ship molds to you from our community makerspace in Kansas City, Missouri. Get in touch at hammerspaceworkshop.com, or submit a job request directly to our armorers at hammerspacehobby.com/forge. ◑

VIRTUAL SAWDUST

Traditional woodworkers can benefit from designing in CAD. Here's how.

Written by Aaron Dietzen

In principle there seems to be a vast difference between woodworking and 3D modeling. Woodworking uses manual tools and natural materials to create physical items that you can touch, feel, and use; 3D modeling uses computers and screens to make virtual designs that only exist in the digital realm. They can be seen as opposites, but used the right way, modeling in 3D offers advantages from which even the most seasoned woodworker can benefit.

I have been modeling in 3D for over two decades. I have designed everything from jewelry to machine parts to entire buildings, and I have seen the clarity that comes to a project when a good 3D model is used to kick off the design process. Full disclosure, I do work for SketchUp (I'm the SketchUp guy on YouTube), but these tips can be leveraged with other software as well. With that, here are five reasons woodworkers should use 3D modeling in their design process.

MEASURE INFINITE TIMES, CUT INFINITE TIMES

Woodworking often involves trying things out several times before it works. That's part of the fun: making something new and figuring out how to make it work. Unfortunately, this can be a time- and material-consuming process when done in the "real world." This is where 3D modeling can save you time, materials, and heartache. Testing a design, or determining material needs in pixels is far less stressful (and costly) than doing so in the shop (Figure **A**).

3D FOR 2D

This can seem a little counter-intuitive, but 3D can be a great way to generate 2D geometry. CNC is a big piece of a modern woodworker's workflow. While toolpaths are generally sent to the machine as 2D geometry, the final pieces need to take into consideration things like material thickness and clearance. Rather than estimate clearances and then find out you were *just* off, model your pieces as components digitally for more accurate dimensions. This will allow you to have the pieces interact as they would in the real world before you purchase material for the CNC (Figure **B**).

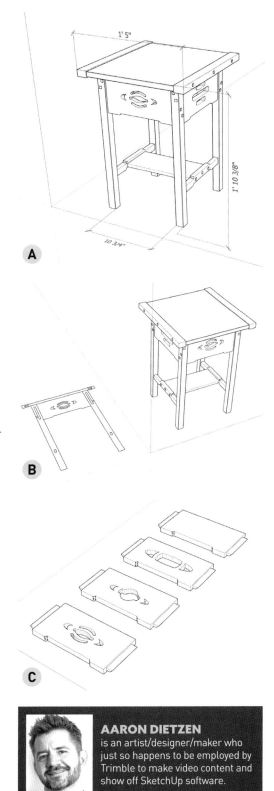

A

B

C

AARON DIETZEN
is an artist/designer/maker who just so happens to be employed by Trimble to make video content and show off SketchUp software.

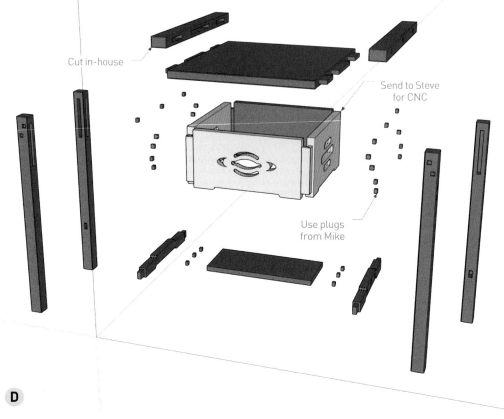

Cut in-house

Send to Steve
for CNC

Use plugs
from Mike

D

SIMPLE DESIGN ITERATION

As you make changes, optimize materials, perfect your design, a 3D model will allow you to simplify tracking your progress. Unlike the storage space needed to keep examples of six iterations of a chair design in the real world, a 3D model will allow you to keep each and every change you make in the same file! This can also be a huge help if you ever want to go back to a previous design choice, or just review the evolution of your design process (Figure **C** on the previous page).

COLLABORATION

One of the best parts of designing in the virtual world is that it is so easy to share information with others. Want to get a friend's input on a table you are considering building? Want to have a coworker help you figure out how to make a stool more stable? What happens if your collaborator lives 1,000 miles away? If you have a 3D model you can just send them a copy of your design. SketchUp and many other CAD programs offer free web versions so collaborators can easily

Dave Richards, Aaron Dietzen

E

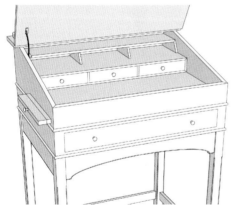

open your model and see everything you need them to see (Figure **D**).

PRE-VISUALIZATION

Possibly the most amazing part of having a 3D model is that you can see exactly how it will look before it exists. And, I am not just talking about seeing the 3D model on a computer, either. Most 3D modelers today have the ability to visualize 3D models in augmented reality. This means you can use your smartphone to see exactly what your project is going to look like when it is complete and sitting in your living room, before you even step foot in your shop. Not only is that satisfying for you, but it makes getting sign off from a significant other much easier when planning your next project (Figure **E**).

From these perspectives, woodworking and 3D modeling are really not that different. 3D modeling helps you get your woodworking project across the finish line. In my opinion, it truly is the best tool for your shop. ⏀

COPPER ELECTROFORMING KIT

$155–$250 enchantedleaves.com

You may know *electroplating* — a thin coat of metal deposited on an object by charging it electrically so it attracts metal ions. *Electroforming* is the same, just thicker — creating a sturdy metal object using the original as a form or scaffold. We spied this DIY kit at Maker Faire last year and it doesn't disappoint — extremely well thought-out and well documented, with all you need to succeed: conductive paint to make your object attractive to copper ions; copper sulfate solution to donate those ions; little details like sealants, adhesives, and jewelry findings; and an optional custom power supply to provide optimal amperage to your electrolyte beaker. So metal! —*Keith Hammond*

CONTROLEO3 REFLOW OVEN BUILD KIT

$249 whizoo.com

If you've taken the plunge into using small SMD components in your electronics projects and you're tired of soldering pins by hand or using a hot air station, you may be longing after a reflow oven like the professionals use. Converting a kitchen toaster oven into a reflow oven is a great option for the serious DIYer, and the Controleo3 Build Kit from Whizoo is the consummate way to confidently compile a complete chip cooking contraption.

The Controleo3 is the controller board that drives the heating elements of the oven. It's powered by open source software, and the included touch screen makes it very easy to navigate the menus and settings. The kit includes everything you need to completely convert a toaster oven into a reflow oven and a very extensive guide with abundant hi-res reference photos. I only wish that the guide listed all the tools you need at the beginning. You will need a lot of different drill bits! Notably, the build kit does not include a toaster oven (I recommend a model similar to the one used in the guide). The whole conversion process took about 12 hours spread over several evenings.

After programming in the reflow profile for my solder paste, I popped several boards into the oven and was astonished when they came out looking professionally assembled. See my build log and boards at projects. jazzychad.net/oven. —*Chad Etzel*

GLASS CUTTING TOOL WITH OIL RESERVOIR

$12 amzn.to/3ihuCJq

For stained glass or making your own picture frames, you need to cut some glass. The vintage style of the tools used for this task has a tiny cutting wheel at the tip of a simple, contoured, pencil-like instrument. Proper use entails the operator dipping the tip into cutting oil before scoring the glass.

In recent years, however, an updated style has become popular. Its general shape and usage remains the same, but the handle is very slightly larger, as it holds an internal reservoir for the cutting oil. Press the tip to score the glass and the oil flows onto the wheel, saving you from having to re-apply during a job. Easy! (Be sure to keep it in the plastic case when not in use, as lighter oils can seep a bit.)

I bought mine, a carbide-tipped version, on Amazon for just over $10, and it's been a useful addition to my toolbox — I mostly use it for cutting bottles. As with any tool, however, you'll get the best results with good technique, so be sure to practice on glass scraps and spare pieces before diving into your masterpiece.
—*Mike Senese*

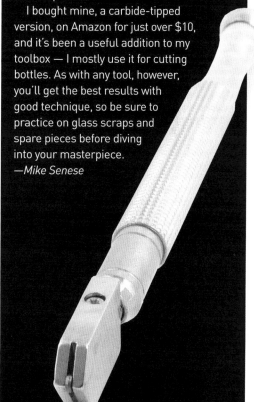

Chad Etzel

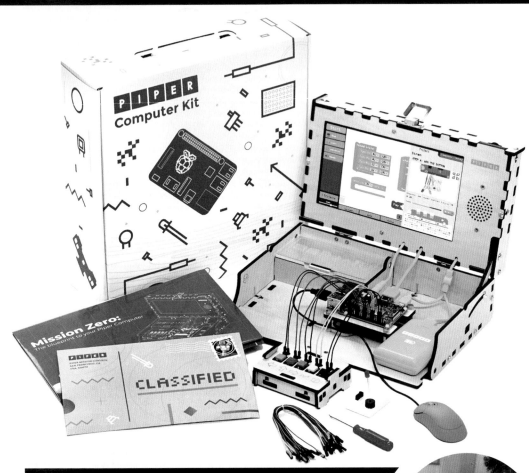

PIPER COMPUTER KIT

$249 playpiper.com

My kids want their own computers and they love building kits, so I was thrilled to see a workshop full of families building this Raspberry Pi-based kit at Maker Faire. At home, my 9-year-old son and I had fun following the giant blueprint to assemble the laser-cut case and connect the peripheral display and speaker. But this kit does a lot more. It's programmed with a Minecraft-based "mission" that requires you to build simple pushbutton circuits on a breadboard to solve challenges within the game, then use those buttons for game play.

The new, improved version adds a bigger LCD, built-in speaker with volume pot, rock-solid mounting for the Pi, and Blockly-based STEAM and game programming projects. There's also a new Arduino-based game controller kit ($59) and sensor kit ($49) to keep the experiments rolling. My kid built the controller one evening before I even finished the dishes, then used it to beat a level he couldn't master before using just the buttons. In our Covid era of remote learning by Zoom, I'm thankful for hands-on learning — and for another machine to free up the parental work laptops. —*Keith Hammond*

DR. DUINO EXPLORER KIT

$159 drduino.com/make

With sensors, speakers, pushbuttons, potentiometers, switches, lights, displays, BLE, and more, the Explorer Kit from Dr. Duino might be the ultimate Arduino shield. Like many kitchen-sink kits, this comes as a packet of raw components encompassing almost every add-on you'd want. Where it differs, however, is its large PCB, measuring 6"×3¼", which allows you to mount all the components, along with an Arduino Nano (included) or Uno (not included), and gives you convenient access to them all when working on a sketch in the Arduino IDE. Spend a couple hours exercising your soldering skills as you follow the online assembly guide to put everything together, then dig into the website project examples to build some fun and educational projects. Then, anytime inspiration strikes, you've got almost any combination of components ready to test out. —*Mike Senese*

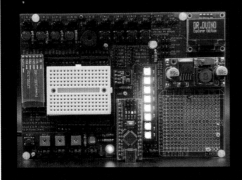

SPEEDBALL DELUXE BLOCK PRINTING KIT

$43 amzn.to/2SaxBsJ

Getting started in any new art medium can be daunting as there are so many materials and tools needed. This kit from Speedball really does well in supplying not only the bare necessities, but also some nice variations to explore. Having never done linocut or block printing, I didn't know what materials I'd like or tools I'd find useful. After a single day with this kit I'm well on my way to building out exactly what I want with confidence.

The only thing missing from the kit was paper, and I have to say it would have been nice to have a few different compositions of paper to learn how they affected the prints. —*Caleb Kraft*

SINGLE-USE SUPER GLUE PACK **$7** amzn.to/3jkWwFL

You're working on a project. You grab the super glue bottle, but can't twist the cap off. You wrench it loose with pliers, but nothing comes out. You jam a thumbtack in the nozzle in hopes of finding a still-liquid reservoir, but it's too late — the glue has all solidified. You throw it in the trash.

After repeating this frustration many times, I spotted these 0.01oz packets at the dollar store. I've successfully reused them, and if one goes bad I've got more in my drawer for a few cents each. Even at online prices, they're a better value than all those wasted bottles. —*Mike Senese*

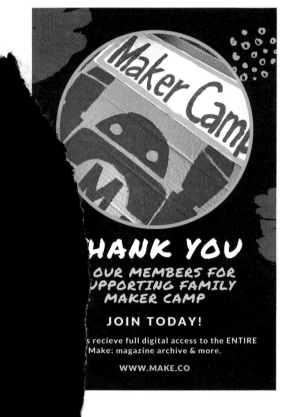

MAD MASK

If you don't cover your nose and mouth, I will gladly do it for you. Written by Allen Pan

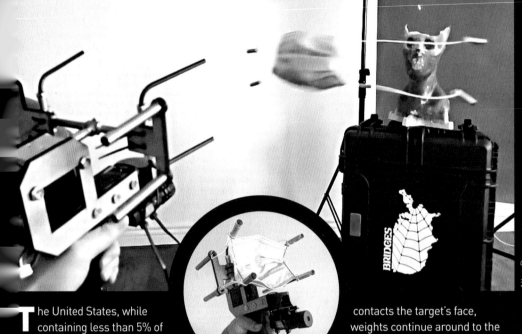

Allen Pan

The United States, while containing less than 5% of the world's population, somehow has roughly 20% of the world's total recorded Covid-19 related deaths. American problems require American solutions, so I built a gun that shoots masks onto people's faces.

Okay, it's not powder actuated, so it's more appropriate (and print-friendly) to call it a mask launcher. A 16-gram CO_2 cartridge powers each "launch" via a high-pressure-rated solenoid valve. The valve outlets to a manifold of four barrels made out of steel brake tubing. Four weighted projectiles fit snugly over those barrels. The weights are connected to high-strength Spectra line, which is fastened to a standard surgical face mask.

To complete the aesthetic, I mounted the entire mechanism on a spray gun grip with a zip-tied green tactical laser meant for a Picatinny rail. It's very patriotic.

When fired, the four weights spread out at a slight angle, keeping the Spectra cord between them taut and the mask open. When the mask contacts the target's face, weights continue around to the back of their head and (with luck) tangle around each other, securing the mask *Alien*-facehugger style.

The response to this project has been overwhelmingly positive, with of course the occasional death threat. Such is internet life. The project seems to really speak to the frustration the majority of people feel when seeing meltdowns and protests over face masks, which are one of our most effective weapons against airborne virus until a vaccine is developed.

NOTE: While hopefully obvious, this mask launcher is just an art piece that is not in any way intended for actual use, and neither I nor *Make:* condone shooting objects or lasers at anyone.

ALLEN PAN sometimes calls himself a failed Mythbuster, "since I didn't quite make the cut in 2017 on the competition reality show *Mythbusters: The Search*." His fallback was a YouTube channel with 1.1 million subscribers: youtube.com/sufficientlyadvanced